Handbook of Machine Tools
Volume 2

HANDBOOK OF MACHINE TOOLS

VOLUME 1   Types of Machines, Forms of Construction and Applications

VOLUME 2   Construction and Mathematical Analysis

VOLUME 3   **Automation and Controls**

VOLUME 4   Metrological Analysis and Performance Tests

Handbook of Machine Tools
Volume 2
Construction and Mathematical Analysis

Manfred Weck
*Lehrstuhl für Werkzeugmaschinen
Laboratorium für Werkzeugmaschinen und Betriebslehre
Aachen, Germany*

Translated from the original German by
H. BIBRING
*Senior Lecturer, Middlesex Polytechnic
London, U.K.*

A Wiley Heyden Publication

JOHN WILEY & SONS
Chichester · New York · Brisbane · Toronto · Singapore

Title of German edition:
Werkzeugmaschinen, Band 1:
Maschinenarten, Bauformen und Anwendungsbereiche
von Prof. Dr.-Ing. Manfred Weck
Laboratorium für Werkzeugmaschinen und Betriebslehre
der Rheinisch-Westfälischen Technischen Hochschule, Aachen
Zweite, neubearbeitete Auflage

© VDI-Verlag GmbH, Düsseldorf, 1980

Copyright © 1984 by Wiley Heyden Ltd.

All rights reserved.

No part of this book may be reproduced by any means, nor transmitted, nor translated into a machine language without the written permission of the publisher.

Library of Congress Cataloging in Publication Data

Weck, Manfred, 1937–
  Handbook of machine tools.

  Translation of: Werkzeugmaschinen.
  'A Wiley–Heyden publication.'
  Includes bibliographies and indexes.
  Contents: v. 1. Types of machines, forms of construction, and applications—v. 2. Construction and mathematical analysis—v. 3. Automation and controls—
[etc.]
  1. Machine-tools.   I. Title.
TJ1185.W382313   1984   621.9'02   83-23483
ISBN 0 471 26226 9 (set)
ISBN 0 471 26223 4 (Vol. 2)

British Library Cataloguing in Publication Data

Weck, Manfred
  Handbook of machine tools.—(A Wiley Heyden publication)
  1. Machine-tools
  I. Title   II. Bibring, H.
  III. Werkzeugmaschinen. *English*
  621.9'02   TJ1185

  ISBN 0 471 26226 9 (set)
  ISBN 0 471 26223 4 (Vol. 2)

Set in VIP Times by Preface Ltd, Salisbury, Wilts.
and printed in Great Britain by Page Brothers (Norwich) Ltd

CONTENTS

Foreword xi

Preface xiii

Nomenclature and Abbreviations xvii

1 Introduction 1

2 Machine Frames and Frame Components 6
   2.1   Requirements of machine-tool frames 6
   2.2   Materials for frame components 6
   2.3   Static loading criteria 9
       2.3.1   Static loads 9
       2.3.2   Static quantifying factors 9
       2.3.3   Analysis of force flux and deformation . . . 11
       2.3.4   Structural design considerations 13
              2.3.4.1   Cross-sections resistant to bending . 14
              2.3.4.2   Torque-resistant cross-sections . . 14
              2.3.4.3   Ribbing 15
              2.3.4.4   Apertures 21
              2.3.4.5   Force transmission 22
              2.3.4.6   Joining methods 24
       2.3.5   Design examples 28
   2.4   Design and shape criteria for dynamic loading . . . 30
       2.4.1   Dynamic loads 30
       2.4.2   Dynamic quantifying factors 32
       2.4.3   Structural design considerations 33
              2.4.3.1   Damping in frames 35
              2.4.3.2   Auxiliary vibration absorbers, friction dampers 37
   2.5   Thermal loading criteria 39
       2.5.1   Thermal loading 39
       2.5.2   Thermal quantifying factors 41
       2.5.3   Structural design considerations 42
   2.6   Mathematical analysis of machine-tool components . . 43
       2.6.1   Fundamentals 43

	2.6.2	Fundamentals of the finite element method	46
		2.6.2.1 Derivation of a matrix for the rigidity of an element	47
		2.6.2.2 Superimposing stiffness matrices of elements to obtain the complete stiffness matrix	50
	2.6.3	Survey of calculations made possible by using the finite element method	53
	2.6.4	Calculation examples	55
		2.6.4.1 Calculations for the static behaviour of machine-frame components	55
		2.6.4.2 Calculations for the dynamic behaviour of machine-frame components	56
		2.6.4.3 Calculations for the thermal behaviour of machine-frame components	59
2.7	Noise reduction in machine design		62
	2.7.1	Fundamentals	63
	2.7.2	Examples of noise reduction	68
		2.7.2.1 Active–primary measures	70
		2.7.2.2 Active–secondary measures	72
		2.7.2.3 Passive–primary measures	72

3 Installation and Foundations of Machine Tools — 74

3.1	Installation and foundations of machines with adequate inherent rigidity	76
3.2	Installation and foundations of metal-forming machines	78
3.3	Foundations for precision machines without adequate inherent rigidity	84
3.4	Foundations for medium and heavy machine tools without adequate inherent rigidity	85

4 Guideways and Bearings — 88

4.1	Hydrodynamic plain guideways and bearings			92
	4.1.1	Fundamentals of friction and lubrication		92
		4.1.1.1	The concept of viscosity	92
		4.1.1.2	Hydrodynamic pressure build-up	94
		4.1.1.3	Types of friction	102
		4.1.1.4	The Stribeck curve	103
		4.1.1.5	Stick-slip effect	104
		4.1.1.6	Raw material pairings and wear factors	106
	4.1.2	Hydrodynamic slideways		111
		4.1.2.1	Guideway elements and design features	111
		4.1.2.2	Clamping devices	117
		4.1.2.3	Compensation for guiding errors	119

		4.1.2.4	Static and dynamic behaviour	120
	4.1.3	Hydrodynamic plain circular bearings		121
		4.1.3.1	Pressure build-up and acceleration characteristics	122
		4.1.3.2	Design variations	125
		4.1.3.3	Hydrodynamic spindle bearing units in machine tools	127
4.2	Hydrostatic plain bearings and guideways			129
	4.2.1	Fundamentals and basic operating principles		129
		4.2.1.1	Oil supply systems	133
		4.2.1.2	Bearing calculations	137
		4.2.1.3	Dynamic behaviour	145
		4.2.1.4	Energy consumption and hydraulic circuits	148
	4.2.2	Hydrodynamic linear bearings		150
		4.2.2.1	Design features and general arrangements	151
		4.2.2.2	Application principles	153
		4.2.2.3	Compensations for guide inaccuracies	156
	4.2.3	Hydrostatic plain circular bearings		157
		4.2.3.1	Design variations	157
		4.2.3.2	Pressure build-up	159
		4.2.3.3	Bearing design	160
		4.2.3.4	Seals and sealing	163
	4.2.4	Hydrostatic spindle bearing systems in machine tools		165
	4.2.5	Hydrostatic lead-screws and nuts		165
4.3	Aerodynamic and aerostatic slideways and bearings			167
	4.3.1	Fundamentals and functional principles		168
	4.3.2	Characteristics		169
		4.3.2.1	Air consumption and load-carrying capacity	169
		4.3.2.2	Dynamic behaviour	170
	4.3.3	Application examples		172
4.4	Rolling guides and bearings			173
	4.4.1	Rolling guides and guidelines		173
		4.4.1.1	Principles of construction	173
		4.4.1.2	Types of design	174
		4.4.1.3	Pre-loading	174
	4.4.2	Rolling cylindrical bearings		176
		4.4.2.1	Survey of bearing designs	176
		4.4.2.2	Bearings for machine spindles and tolerances for the associated shafts and bearing seatings	177
		4.4.2.3	Bearing play	178

		4.4.2.4	Resilience and pre-loading of radial bearings	179
		4.4.2.5	Resilience and pre-loading of thrust bearings	183
		4.4.2.6	Comparison between radial and axial resilience curves for various bearings .	184
		4.4.2.7	Cage slip on radial bearings . . .	185
		4.4.2.8	Vibration excitations from rolling bearings	185
		4.4.2.9	Lubrication and temperature effects .	187
		4.4.2.10	Characteristics of rolling bearings compared with other bearings . . .	189
	4.4.3	Spindle bearing units employing rolling bearings for machine-tool construction		190
		4.4.3.1	Specifications and design principles .	190
		4.4.3.2	Static behaviour	193
		4.4.3.3	Dynamic behaviour	198
		4.4.3.4	Thermal behaviour	201
	4.4.4	Recirculating ball spindles and nuts . . .		203

5* Main Drives 206
5.1 Motors 206
5.1.1 Electrical machines 207
5.1.1.1 Direct current (DC) machines . . 207
5.1.1.2 Synchronous machines . . 214
5.1.1.3 Asynchronous (induction) machines . 216
5.1.2 Hydraulic motors 220
5.1.2.1 Rotary hydraulic drives . . . 223
5.1.2.2 Linear hydraulic drives . . . 228
5.1.2.3 Speed control 229
5.1.3 Start-up conditions of a drive 233
5.2 Transmission drives 235
5.2.1 General requirements 235
5.2.2 Uniform transmission drives 236
5.2.2.1 Drives with stepped-speed changes . 236
5.2.2.2 Drives with stepless-speed changes . 249
5.2.3 Non-uniform transmission drives . . . 265
5.2.3.1 Slotted-link mechanism . . . 266
5.2.3.2 Crank mechanism 266
5.2.3.3 Crank-rocker mechanism . . . 268
5.2.3.4 Cams 269
5.3 Couplings and clutches 269
5.3.1 General requirements 269

* Feed mechanisms, see Volume 3.

	5.3.2	Permanent couplings		270
		5.3.2.1 Rigid couplings		271
		5.3.2.2 Flexible couplings		271
	5.3.3	Clutches		274
		5.3.3.1 Externally operated clutches		274
		5.3.3.2 Self-acting clutches		282

6 Summary 286

7 References 287

8 Index 291

FOREWORD

Machine tools are among the most important means of production for the metal-working industries. Without the development of this type of machine, the high living standards of the present time would be unthinkable. In some of the most highly industrialized nations, approximately 10% of all machines built are machine tools, and about 10% of the work force in machine manufacture is concerned with machine tools.

The form of construction and degree of automation of machine tools are just as varied as their fields of application. The wide field that is embraced ranges from casting and forming, through cutting (including abrasion), to welding and assembly, and is limited only by the degree of technological development. According to the component to be produced and the quantities involved, these machines are variedly automated with a greater or lesser degree of flexibility. Thus, single-purpose or special-purpose machines are available to the user, as are universal machines offering a wide range of applications.

As a result of the increased demands in both performance and precision, the manufacturer of these machines has to determine the optimum design of the individual machine components. To this end, he requires an in-depth knowledge of the capabilities and capacities of the components and elements of his machine.

Today, as a result of the availability of a very comprehensive library of programs, the necessary calculations can be carried out with the aid of computers. Metrological analysis and objective performance tests have opened up the possibilities of critically determining the machine's capabilities and degree of accuracy of production, as well as determining its geometric, kinematic, static, dynamic, thermal and acoustic properties, and from this information, to initiate any improvements thought to be necessary.

The constant tendency towards further automation of machine tools has resulted in the development of a wide range of alternative controls. In recent years, the development of electronics has had a marked effect on machine-tool controls. Microprocessors and process calculators have made possible control techniques which were previously unthinkable. Mechanization and automation have also been applied in the area of material transportation and the feeding of the machines with material. The work done in these fields has resulted in the availability of transfer machines for the mass production industry and for medium- and small-quantity manufacture, and has led to the availability of flexible production systems.

The four volumes published under the general title '*Handbook of Machine Tools*' are intended for the use of students in the field of production engineering, as well as established specialists in the field who require to keep abreast of the constantly developing field of this branch of machine construction. A further aim of these volumes is to assist users in the choice of suitable machines and their controls. For the machine-tool manufacturer, optimum layouts of machine components, drives and controls are indicated, and it is shown how improvements can be achieved as a result of metrological analysis and objective performance tests. Their content is derived from and well validated by lectures given at the Technical University for the Rhineland and Westphalia in Aachen.

MANFRED WECK

PREFACE

The efficient functioning of a machine is highly dependent upon a carefully designed layout of its components and elements which will also eliminate time-wasting subsequent modifications and adaptations. This is particularly so in the case of complex machine tools, from which high-performance standards are demanded, with regard to accuracy under static, dynamic and thermal loading conditions.

The aim of this volume is to provide the machine builder with the information necessary for the design and mathematical analysis of machine-tool elements and their structural components in accordance with the most up-to-date knowledge of this technology. When the components of a machine tool are scientifically designed, the accuracy and performance of the machine are improved, having a marked influence on the quality and cost of the products to be produced.

The following are dealt with in detail:
  Machine frames and frame components, machine beds and foundations, spindle bearings and guiding systems, motors for main drives, stepless and stepped-drive mechanisms and clutches

Feed drives are covered in detail in Volume 3 where they are described in connection with and from the point of view of control technology.

In all cases the mathematical analysis is detailed as, for example, the establishment of static, dynamic and thermal relationships. The construction details are discussed both from the traditional as well as from newer concepts.

The preparation of this volume was greatly facilitated through the help of my colleagues: Dipl.-Ing. N. Diekhans, Dr.-Ing. M. Gather, Dipl.-Ing. A. Gohritz, Dipl.-Ing. H. Heinrichs, Dipl.-Ing. K. Klumpers, Dipl.-Ing. A. Kruse, Dipl.-Ing. B. Neupert and Dipl.-Ing. H. Peuler. The co-ordination of the separate chapters of this volume was undertaken by Dipl.-Ing. B. Thurat. I would like to take this opportunity of expressing my gratitude to these gentlemen for their readiness to assist in this work. I am also indebted to Dr.-Ing. W. Borchert of VDI-Verlag for the painstaking examination of the manuscripts and to A. Hümmler, who took great care in proofreading the English manuscript.

MANFRED WECK

NOMENCLATURE AND ABBREVIATIONS

Upper case letters

A	mm²	Area
A	mm	Axle centres
A_R	mm²	Friction area
B	mm	Width, breadth
B	—	Speed ratio
B_d	—	Diameter ratio
B_0	—	Speed ratio of stepless drive
B_{st}	—	Speed ratio of stepped drive
B_v	—	Velocity ratio
C	N	Centrifugal force
C	—	Motor constant
D	mm	Length
D	mm	Outside diameter of bearing
D	—	Damping ratio
D_W	mm	Diameter of a rolling element
E	N mm^{-2}	Modulus of elasticity
F	N	Force
F_F	N	Force on guide
F_M	N	Force due to mass
F_P	N	Force in piston rod (connecting rod)
F_R	N	Frictional force
F_S	N	Press force, force in direction of stroke
F_Z	N	Load on each rolling element
G	μm N^{-1}	Flexibility
$G(j\omega)$	μm N^{-1}	Flexibility frequency characteristic
I	A	Current
I_A	A	Armature current
I_P	mm⁴	Polar second moment of area
I_T	mm⁴	Torque resistance
I_x, I_y	mm⁴	Axial second moment of area
J	kg m²	Moment of inertia
K	—	Tremor factor, K value
K	N μm^{-1}	Stiffness
L	mm	Length
L	dB	Sound level
L_{WA}	dB	Acoustic-power level

Symbol	Units	Description
M	—	Bearing constant
P	W	Power
P	W	Acoustic power
P_{Mot}	W	Motor power
P_R	W	Frictional energy
Q	cm³ s⁻¹	Flow quantity
Q_{Mot}	cm³ s⁻¹	Hydraulic motor-suction flow quantity
Q_P	cm³ s⁻¹	Hydraulic pump-delivery flow quantity
Q_s	cm³ s⁻¹	Oil-supply flow quantity
R	Ω	Resistance
R	N s cm⁻⁵	Hydraulic resistance
R	dB	Sound-barrier ratio
R	—	Number of subdivisions within a power of ten
R_F	N s m⁻³	Radiation factor
$R_{A\,Gen}$	Ω	Armature resistance of generator
$R_{A\,Mot}$	Ω	Armature resistance of motor
S	m²	Sound-radiation area
T	K	Temperature
T	s	Time constant
T	N m	Torque
T_B	N m	Acceleration torque
T_E	N m	Engagement torque
T_K	—	Transmission ratio
T_{Load}	N m	Torque load
T_{Nil}	N m	No-load torque
T_{Mot}	N m	Motor torque
T_T	N m	Transmitted torque
U	mm	Displacement at one node
$\dot{U}$	mm s⁻¹	Velocity at one node
$\ddot{U}$	mm s⁻²	Acceleration at one node
V	V	Voltage
V	cm³	Suction or delivery volume
V	cm³	Volume
$V_{A\,Gen}$	V	Rotor voltage of generator
$V_{A\,Mot}$	V	Rotor voltage of motor
V_{Mot}	cm³	Motor-suction volume
V_P	cm³	Pump-delivery volume
W_A	W	External work done
W_B	W	Work done in acceleration
W_{Frict}	W	Work done by friction
W_I	W	Internal work done
W_{tot}	W	Total work done
Z	N s m⁻¹	Impedance

Lower case letters

Symbol	Units	Description
a	mm	Vane thickness

a	mm	Cantilever length
a	m s^{-2}	Acceleration
b	mm	Width, breadth
b	mm	Distance between bearings
c	m s^{-1}	Sound velocity
c	—	Arithmetic progression speed step
c	N s cm^{-1}	Damping constant
d	μm N^{-1}	Flexibility
d	mm	Diameter
d_m	mm	Mean bearing diameter $(D+d)/2$
d_{max}	mm	Maximum workpiece diameter
d_{min}	mm	Minimum workpiece diameter
d_{01}	mm	Pitch circle diameter – gear 1
d_{02}	mm	Pitch circle diameter – gear 2
e	mm	Eccentricity
f	Hz	Frequency
f_E	Hz	Effective frequency
h	mm	Height
h	mm	Length of stroke
h	mm	Height of gap
i	—	Transmission ratio
i	A	Rectifier current
i	—	Number of rows of rolling elements
k	N μm^{-1}	Stiffness
k	—	Position of a preferred number in a preferred number series
k	—	Correction factor
k	—	Overlap ratio
k	—	Constants
k_A	N μm^{-1}	Stiffness of front bearing
k_B	N μm^{-1}	Stiffness of rear bearing
k_F	N μm^{-1}	Spring stiffness
k_K	N μm^{-1}	Stiffness at point-of-force application due to stiffness of headstock
k_L	N μm^{-1}	Stiffness at point-of-force application due to stiffness of bearing
k_{sp}	N μm^{-1}	Stiffness at point-of-force application due to stiffness of spindle
l	mm	Length of outflow
l	mm	Length of wedge gap
l_a	mm	Loaded length of rolling element
m	—	Surface-finish index
m	mm	Module
m	kg	Mass of a body
m'	—	Reduction in wedge gap
n	rev min^{-1}	Speed, r.p.m.

n_p	rev min^{-1}		Speed of pump
n_{0Pr}	rev min^{-1}		No-load speed with primary speed regulation
n_{0Sr}	rev min^{-1}		No-load speed with secondary speed regulation
p	%		Reduction in cutting velocity
p	N mm^{-2}		Pressure per unit area
p	bar		Pressure
p	—		Number of pole pairs
p_P	bar		Pump-oil pressure
r	mm		Radius
s	mm		Distance
s	mm		Wall thickness
s	—		Slip
t	s		Time
t_F	s		Friction time
v	m s^{-1}		Velocity
v_1	m s^{-1}		Lower cutting velocity
v_u	m s^{-1}		Upper cutting velocity
v_s	m s^{-1}		Flow velocity
x	—		Nominal number, standard number
y	μm		Total radial spindle displacement
y_L	μm		Radial spindle displacement due to flexibility of the bearing (bearing contribution)
y_k	μm		Radial spindle displacement due to flexibility of the bearing mounting components (spindle casing, headstock)
y_{sp}	μm		Radial spindle displacement due to flexibility of the spindle (spindle contribution)
y_1, y_2	—		Nominal or standard auxiliary quantities or variable
z	—		Number of cells
z	—		Number of pistons, vanes or teeth
z	—		Number of speed steps
z	—		Number of rolling elements per row
z_b	mm		Width of teeth

Greek letters

α	degree	Angle
α	degree	Contact angle
α	degree	Swash-plate set angle
α	μm m^{-1}	Angle of tilt
α	—	Surface-finish characteristic
β	—	Resilience ratio (coefficient)
ΔL	dB	Noise-transmission level difference
Δn	rev min^{-1}	Speed reduction
Δn_{pr}	rev min^{-1}	Speed reduction after primary control

Δn_{se}	rev min^{-1}	Speed reduction after secondary control
δ	μm	Bearing resilience
δ	μm	Resilience path
δ_e	μm	Elastic component of resilience
δ^*	μm	Resilience of play-free bearing
δ_{max}	μm	Resilience of supporting roller ($\psi = 0$)
ε	—	Pump efficiency
ε	—	Expansion
η	N s cm^{-2}	Dynamic viscosity
θ	K	Temperature
θ_0	K	Temperature distribution
κ	—	Resistance ratio R_{a0}/R_{u0}
λ	m	Wavelength
μ	—	Mass ratio
μ	—	Coefficient of friction
ν	cm^2 s^{-1}	Kinematic viscosity
ν	—	Resonance peak
ξ	—	Resistance ratio R_K/R_c
ρ	g mm^{-2}	Density
σ	—	Radiation rate
σ	N nm^{-2}	Normal stress
τ	s	Time constant
τ	N mm^{-2}	Shear stress
Φ	V s	Magnetic flux
ϕ	V$_s$	Geometric progression step
ϕ	degree	Angle of rotation
ψ	—	Relative displacement
ψ_0	degree	Angle of zone under load
Ω_0	l s^{-1}	Natural frequency
ω	l s^{-1}	Angular velocity
ω	l s^{-1}	Angular velocity of shaft
ω_m	l s^{-1}	Angular velocity of rolling-element assembly
ω_{WP}	l s^{-1}	Angular velocity at working point
$\omega_{1/2}$	l s^{-1}	Angular velocity, input/output

Indices, Suffixes

A	Armature, rotor
a	Axial
B	Acceleration
c	Cell
dyn	Dynamic
e, eff	Effective
f	Excitation

Gen	Generator
i	Induced
K	Capillary
L	Bearing, load
loss	Loss
max	Maximum
mech	Mechanical
min	Minimum
Mot, M	Motor
N	Relative to nominal state
P	Pump, Piston
pull-out	Relative to pull-out state
r	Radial
Start	Relative to starting or acceleration drive
s	Synchronized
stat	Static
th	Theoretical
tot	Total
u	Circumferential
v	Before
Val	Valve
WP	Working point
0	No-load state, free-wheeling

Vectors and Matrices

$\{\ \}$	Vector
$\{\ \}^T$	Transposed vector
$[\ ]$	Matrix
$[\ ]^T$	Transposed matrix

1

INTRODUCTION

The accuracy, performance, capability and environmental effects of a production unit very largely affect the quality of the goods being produced, as well as the ability to utilize the machine for economic productivity.

The designer must be able to guarantee that the machine will be capable of meeting all the demands which may be expected as a result of the conditions under which it will be used.

The design cycle is broadly illustrated on the left side of Fig. 1.1. The individual stages, apart from any preconceived specification, consist of development, design and calculations, as well as the preparation of detailed drawings and the provision of the production equipment. A variety of aids are available for these stages in the cycle. The following are used in the design and calculation stage:

(a) computer software for the calculation of the design details of the machine components and the complete machine;
(b) metrological and analytical techniques to determine the actual relationships of the machine parts (Volume 4);
(c) design indexes and standards as references for simplified solutions in relation to the main construction of a given requirement;
(d) standards, which determine the acceptability of particular machine characteristics (Volume 4).

According to the type and complexity of a design problem, the designer today has mini or desk computers available to help solve routine calculations, as well as large computers for more complicated problems.

The designer can draw to an increasing extent on recorded results of single and series of tests and measurements which indicate the current standing of the technology; these may then be used for comparison when making an appraisal of a machine, and form a basis for decision making (Volume 4). The results of a large number of tests and measurements taken on previously built machines, or mock-ups, as well as theoretical studies utilizing computer models and a wide range of mathematical solutions, are recorded in increasing

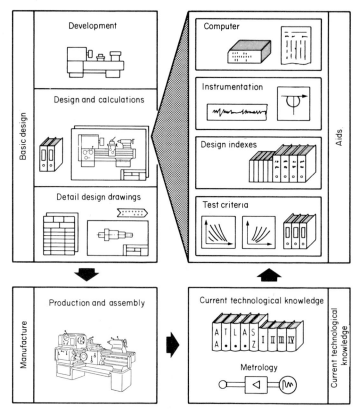

Fig. 1.1 Block diagram of a machine design cycle

numbers in so-called design indexes. The designer can adapt this recorded build-up of knowledge in order to apply the rules laid down for his own specialized requirements.

If a particular design cannot be evaluated with a reasonable number of direct measurements, then a range of performance tests and aids are available, e.g. mechanical efficiency analysis or appropriate acceptance charts and diagrams, which enable speedy, but reliable, decisions to be made. This type of aid must reflect the current knowledge of technological development and must be continuously up-dated.

The relevant criteria for the evaluation of a machine tool are presented in the columns of the table in Fig. 1.2. The accuracy of performance and capabilities of the machine are mainly determined by the use of established methods of measurement of kinematic, static, dynamic, thermoelastic and stress characteristics. (The control aspects and the influence of the degree of automation on the economic utilization of the machine are dealt with in Volume 3.)

The success of the machine in the market-place today is also increasingly

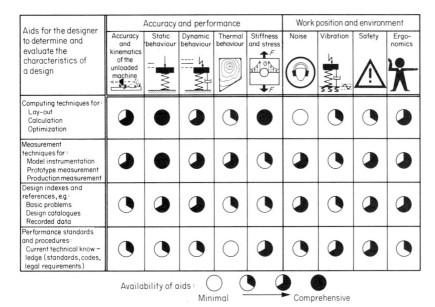

Fig. 1.2 Availability of aids for machine design

dependent upon its acceptability in relation to the environment and the place in which it is to be installed. Thus there is concern about the generation of noise and vibrations on the one hand and the safe operation and ergonomics of the machine on the other. The information and data available for this design area are indicated in the table. When these are listed under individual sections of the various criteria, it is apparent that their availability is quite varied. However, when taken as a whole, such information and data combine to have a greater and increasing influence on the final design.

In particular, machine-tool construction which occupies such a key position demands the adoption of modern technology and aids in the development and design stage. The market position requires particularly that a flexible approach be adopted to the necessary production facilities in order to cope with the constantly changing situation.

A crucial aid for design is the electronic data processor, generally known as a computer, with its varied peripheral devices. A prerequisite for the effective use of a computer is the availability of appropriate and useful user's software, i.e. computer programs.

When one examines from the user's point of view the programs which are available today in the field of machine-tool construction, it becomes clear that they may be classified into three groups, as may be seen in Fig. 1.3.[1,2] The 'component-orientated' programs shown on the left are concerned with particular components or assemblies, e.g. the calculations of spindle/bearing systems, the selection and calculations for gear drives, bearings, etc. Such programs are entirely dependent on the component for their data input,

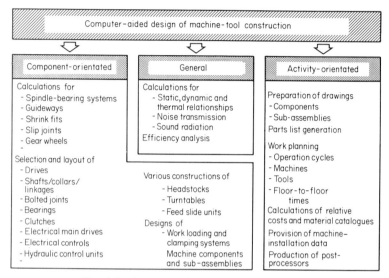

Fig. 1.3 Programs for computer aided design

calculations and output. This is particularly the case for programs for unit-construction design, which the user must adapt to suit his specialized specification.

The 'general' programs listed in the centre are in quite a different category. They are universally applicable, but generally require a more detailed data preparation and input, as well as output evaluation.

The 'activity-orientated' programs have a wide field of application, and may be used for automating the 'simpler' design activities. Applications for such programs are, for example, the production of drawings for individual components and assemblies, the generation of parts lists, the planning of the work programme and control programs for numerical control (NC) working. (See *Organisation for Production Technology* by Walter Eversheim.)

Software for an appropriate computer may be found useful, even during the early development stages. Conversely, evaluations requiring measurement require a finished physical object, which may be a model, prototype or even a production machine. All essential judgements are dependent upon an analysis of measured data. However, the costs involved are often considerable (Volume 4).

A further design aid which is of value in the construction and calculation stage is the design index. The designer uses it in the manner of a reference work. The design index contains solutions to design problems listed in accordance with technical and economic criteria. As indicated in Fig. 1.4, these may be established by a systematic examination either of the variables based on theoretical calculations or of the data collected from suitably instrumented measurements of models. Moreover, results from practical examples are evaluated, systematically listed and similarly collated in design indexes. The

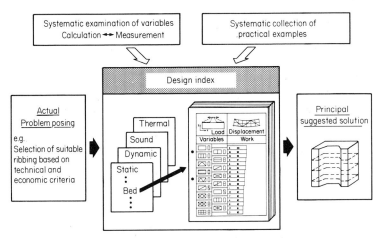

Fig. 1.4 Development and utilization of design indexes

layout of such design indexes must be so arranged that a principal solution may be found for a given problem in the appropriate section. Gradually, this information enables the final design to evolve, and, by the observance of some rules provided, it can also be fully dimensioned.

In this second volume *Construction and Mathematical Analysis* of the compendium *Machine Tools*, the basic relationship between a mathematical analysis of the individual machine components and their main physical characteristics is described and amplified with examples. A systematic comparison and discussion of alternative solutions is intended to provide the designer with an impetus to find suitable answers to his own, constantly changing, specialized problems. As the necessary mathematics are usually highly complex, there is available today—as mentioned earlier—a comprehensive software library, enabling computer-aided designs of machine elements to be completed.[1,2]

2

MACHINE FRAMES AND FRAME COMPONENTS

2.1 Requirements of machine-tool frames

The frames and their components are the load-carrying and supporting bodies of machine tools. They are required to support and guide the individual constructional and functional elements; their size and shape is determined by particular functions of the machine. Their form is basically dependent upon the position and length of the moving axes, and upon the consequential arrangement of the components and sub-assemblies (e.g. work spindles, slides, supports, drives, motors, control units). In addition, they are influenced by the magnitude of the process forces and accessibility for their own construction, as well as their use and operation.

To facilitate their production and assembly, machine frames are frequently constructed from several individual components which are then bolted to each other at the joints. In rare cases adhesives are applied. Figure 2.1 illustrates typical constructional forms of machine tools. Some frequently used structural components are detailed in the illustration.

The masses of the moving components of the machine and the work, as well as the machining forces, must cause only minimal distortions of the machine, whilst the swarf produced must be rapidly removed from the working area. Such requirements have, for example, led to the conception of an inclined bed for a NC chucking-centre lathe shown in Fig. 2.2.

The following section considers the static, dynamic and thermal characteristics from the point of view of machine-tool frame construction.

2.2 Materials for frame components

The materials used for machine frames and frame components are steel, steel castings and cast iron. In some special cases, steel-reinforced concrete is also used.

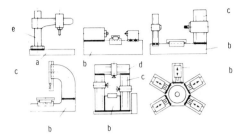

(a) Base plate (b) Bed (c) Column (d) Cross beams
(e) Pillar column

Open 'C' construction		Closed 'O' construction	
Single-frame construction	Twin-frame construction	Double-column construction	Pillar frame construction

Fig. 2.1 Typical structural machine tool components (upper part); constructional designs of press frames (lower part)

Manually operated universal lathe

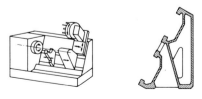

Automatic lathe with inclined bed

Fig. 2.2 Adaptation of a lathe bed for a new machining concept

The material choice is governed by the following properties of materials, which will determine the performance of components:

Strength (elastic limit, fatigue limit)	⟶	Avoidance of plastic deformation and fracture
Specific density	⟶	Mass distribution, static and dynamic performance
Modulus of elasticity, torsion or stiffness modulus	⟶	Static and dynamic performance
Damping properties	⟶	Dynamic performance
Coefficient of friction, hardness	⟶	Frictional and wear performance of moving surfaces
Residual stress, creep, relaxation properties	⟶	Long-term retention of geometric shape
Coefficient of thermal expansion, specific heat, coefficient of thermal conductivity	⟶	Thermoelastic characteristics

Table 2.1 lists some physical-property values for materials widely used in machine-tool construction.

Table 2.1 Physical properties of materials

Material	Modulus of elasticity E(GN m^{-2})	Specific density γ(kN m^{-3})	Coefficient of thermal expansion α(l K^{-1})	Tensile strength σ_B(MN m^{-2})
Steel	210	78.5	11.1×10^{-6}	400–1300
Steel casting	170	74.0	9.5×10^{-6}	400–700
Grey cast iron	50–110	72.0	9.0×10^{-6}	100–300
Copper	120	89.5	16.2×10^{-6}	200–400
Aluminium	70	27.0	23.8×10^{-6}	120–400
Brass	90	85.0	19.0×10^{-6}	300–700
Titanium	110	45.0	10.8×10^{-6}	500–1200
Concrete	20	25.0	11.0×10^{-6}	5–60

Apart from these material properties, the choice is also influenced by production and economic factors. These are in the main as follows:

(a) costs of material;
(b) economic production;
(c) ease of working;
(d) casting or welding properties.

The static quality is mainly dependent upon the modulus of elasticity. Whilst this value may be stated quite precisely for steel, in the case of cast iron the modulus can vary for differing wall thicknesses and loading conditions.

Whenever vibrations are to be considered, the damping properties of the material become important. However, in most metallic materials, these damping properties are so low that they may be ignored when considering the damping conditions which occur at joints and in bearings.

2.3 Static loading criteria

2.3.1 Static loads

The static loads of a machine tool result from the process forces and the masses of the work and machine components. Owing to the changing conditions during machining, the magnitude and direction of the forces and moments change, as well as the point of stress intensity. This results in varying deformations of the frame.

Figure 2.3 gives an example of a typical deformation of a press as a result of an off-centre static load. The linear load causes a twist in the frame and the torque moment tends to bend it.

The static working loads and the mass of the work being machined produce distortions which result in the production of geometric errors on the work in metal-cutting machine tools. Thus the work tolerance which the machine is expected to hold demands adequate stiffness in the frame in the case of cutting and erosion machines, whilst in the case of metal-forming machines the tolerance permitted to the moving tools sets similar requirements.

2.3.2 Static quantifying factors

The static characteristic of a machine tool, its subassemblies or individual components is dependent upon the elastic deformations which occur due to the applied loads at any given time. Hence it becomes clear that the most important evaluation is that of stiffness or rigidity and the reciprocal thereof, i.e. the flexibility.

The relationship between a deformation x and the force F producing it is represented in the form of curves of characteristics. Only in the case of

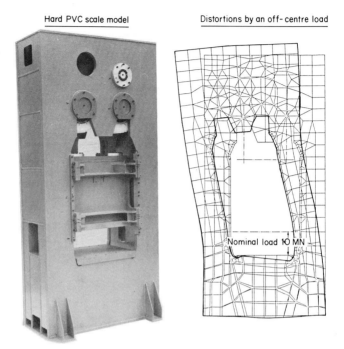

Fig. 2.3 Static deformation of a press frame resulting from an off-centre loading. (Thyssen Industrie AG)

jointless components is such a relationship in linear form. Most of these curves for actual components and machines indicate disproportional increases in distortion. Any increase observed in rigidity may, in the main, be traced back to increases in the active contact areas in the contact zones.

Two relationships for stiffness k at a working point can be evolved (see Fig. 2.4). The first relationship (on the left in the diagram) shows a secant drawn

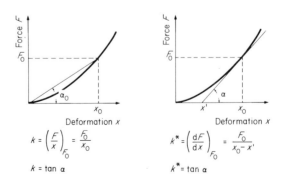

$$k = \left(\frac{F}{x}\right)_{F_0} = \frac{F_0}{x_0}$$

$$k = \tan \alpha$$

$$k^* = \left(\frac{dF}{dx}\right)_{F_0} = \frac{F_0}{x_0 - x'}$$

$$k^* = \tan \alpha$$

Fig. 2.4 Stiffness relationships

from the point under consideration, F_0, x_0, through the origin:

$$k = \left(\frac{F}{x}\right)_{F_0} = \frac{F_0}{x_0} \quad (\text{N } \mu\text{m}^{-1}) \tag{2.1}$$

This relationship is applied when a mean value for stiffness is adequate under a given loading condition ($0 < F < F_0$).

For the second relationship for stiffness (on the right in Fig. 2.4) the slope of the tangent to the curve at the point under consideration, F_0, x_0, is used:

$$k_{(F=F_0)} = \frac{dF}{dx_{(F=F_0)}} = \frac{F_0}{x_0 - x} \quad (\text{N } \mu\text{m}^{-1}) \tag{2.2}$$

This method of calculation permits the deformation, which is the result of an applied force in addition to a preload F_0, to be determined (e.g. dynamic loading of a machine already under stress).

The flexibility d is the reciprocal of stiffness and hence is given as a relationship of the deformation x to the force F:

$$d = \frac{dx}{dF} = \frac{1}{k} \quad (\mu\text{m N}^{-1}) \tag{2.3}$$

The stiffness, and hence the flexibility, are dependent upon the material used, the geometric shape, and the type, position and direction of the applied force on the component. The layout of individual frame components can only be established in relation to all the other elements and subassemblies of the machine. The main consideration is to produce the minimum relative displacement between the work and tool as a result of the applied forces.

2.3.3 Analysis of force flux and deformation

The deformations which occur at the point of load application between the tool and work due to the machining forces are a result of all the force transmissions from all the relevant machine components and constructional units. An analysis of the force flux and deformation pattern examines the stressing of the individual machine elements and their separate contribution to the total displacements at the points of load application.

Figure 2.5 depicts diagrammatically the force flux on a horizontal boring machine. It can be seen that there is a need to ensure adequate stiffness of all the elements which lie in the path of the force flux, including bearings, spindles and slides, so that an acceptable stiffness of the whole machine ensues. A simplified analogous approach regards the machine components as a number of springs connected to each other either in series or in parallel, so that the total flexibility may be derived by the addition of the flexibility of

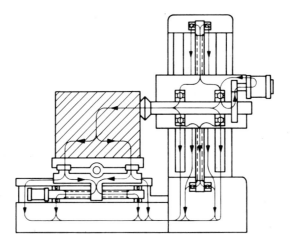

Fig. 2.5 Force-flux diagram of a machine tool

each individual spring:

$$d_{\text{tot}} = \frac{1}{k_{\text{tot}}} = \underbrace{\frac{1}{k_1} + \frac{1}{k_2}}_{\text{connected in series}} + \underbrace{\frac{1}{k_3 \ k_4}}_{\text{connected in parallel}} + \cdots \quad (2.4)$$

Thus the complete machine is always 'softer' than the most resilient machine unit situated in the force-flux flow.

An example of an analysis for the deformation of a horizontal boring machine is shown in Fig. 2.6. The loads $F_x = F_y = F_z = 40{,}000$ N are acting in the respective co-ordinate axes on the spindle when it is in the working position, as indicated on the right part of the diagram. On the left of Fig. 2.6, the separate contributions of the various machine units to the total displacement are indicated and represented in the form of a histogram. The main spindle and its housing are subjected to bending forces in both the x and y directions, and are considerably more resilient in these directions than in the z direction, where a compressive load is acting. Whilst the column exhibits more or less equal properties in all three directions, the deformations of the column and bed are markedly different. The column is subjected to a heavy torque loading due to force F_x about the y axis, and this results in a much higher deflection than produced by the bending loads in the other directions of force application. The displacement at the point of force application, which may be traced back to deformations of the bed, varies a great deal in the different directions due to the large variation of the length of leverage which is possible.

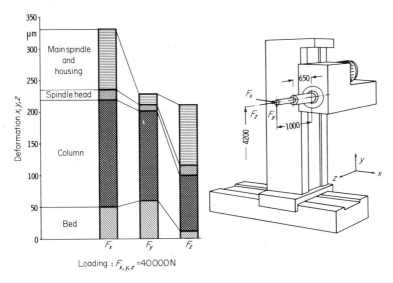

Fig. 2.6 Deformation analysis of a horizontal boring machine

2.3.4 Structural design considerations

The structural design of the sections of the machine components must take into account both bending and torque loading conditions. An evaluation of the static characteristics of a frame component is possible when its geometry is not too complex and when the overriding loading is either bending or torque. This is particularly the case for components which have a single main direction of deformation, e.g. columns, which may be considered as beams. Thus the resistance to bending is mainly dependent upon the second moment of area about the neutral axis, whilst the torsional stiffness is largely governed by the polar moment of area (Fig. 2.7). Moreover, it must be noted that the

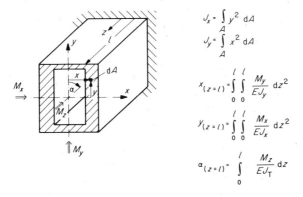

$$J_x = \int_A y^2 \, dA$$
$$J_y = \int_A x^2 \, dA$$
$$x_{(z=l)} = \int_0^l \int_0^l \frac{M_y}{EJ_y} \, dz^2$$
$$y_{(z=l)} = \int_0^l \int_0^l \frac{M_x}{EJ_x} \, dz^2$$
$$\alpha_{(z=l)} = \int_0^l \frac{M_z}{EJ_T} \, dz$$

Fig. 2.7 Second moment of area about the neutral axis

load forces normally do not act along the axis which passes through the centre of gravity of a cross-section, but is frequently introduced along interfaces and guideways in the weaker wall thickness of the frame. Consequently, the way in which the forces are introduced is also significant for the rigidity of a component, as is its constructional layout.

2.3.4.1 Cross-sections resistant to bending

The resistance to bending of a rectangular beam is proportional to its second moment of area about the neutral axis. The equations for these moments about the x and y axes of a cross-section are given in Fig. 2.7. The designer can note from these that thin-walled cross-sections in which the walls are widely spaced apart give high resistance to bending, provided that there are no cross-sectional deformations, indentations, etc. Figure 2.8 illustrates the relative second moment of area about the neutral axis for differing cross-sections.[3] The areas of each of the cross-sections depicted are assumed to be equal.

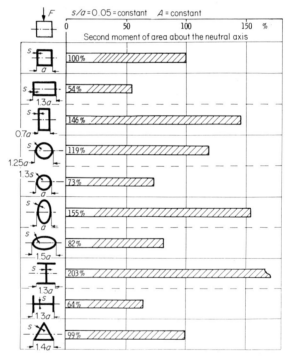

Fig. 2.8 Second moment of area about the neutral axis of differing cross-sections

2.3.4.2 Torque-resistant cross-sections

The torque resistance I_T is a measure of the stiffness of a beam under applied torque loads. The torsional shear stress gradient in a closed cross-section is

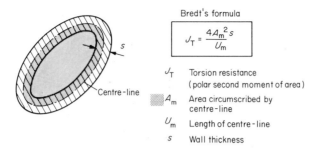

Fig. 2.9 Torsion resistance and Bredt's formula

always linear. Cross-sections which are not in a closed-box form have a reduced torque resistance when compared with closed-box-type cross-sections, as the torsional shear stress is both axial and radial in the outer skin.

The torque resistance I_T cannot be calculated for all cross-sections using the same algorithm. For example, in the case of thick-walled or solid cross-sections, the value equals the polar second moment of area:

$$I_T = I_P = \int_A (x^2 + y^2) dA = I_x + I_y \qquad (2.5)$$

For closed, thin-walled cross-sections, the Bredt formula is applied, as shown in Fig. 2.9. In the case of cross-sections which do not have a closed-box form, the mathematical treatment requires the use of differential equations (potential equations) which have closed similarities to Prandtl's 'soap bubble analogy' and to 'flow equations'.

Figure 2.10 shows a comparative diagram of the torque resistance for some simple cross-sections. It can be seen that the open-I cross-section, which is particularly good in resisting deformation due to bending stresses, has very little torque resistance; the thin-walled circular cross-sections show the best resistance to torsional deformation.

Torque loading of machine-frame components is usually the result of a force couple acting at the guideways. This type of loading causes distortions of the cross-sections, which in turn result in further reductions in torsional stiffness.

Figure 2.11 shows the deformation mechanism causing distortions in the cross-section of a simple machine column which has three couples acting in differing planes. It may be noted from the diagrams that the distortions increase as the distance of the loading point from the support point becomes greater.

2.3.4.3 Ribbing
The bending as well as the torque resistance of machine-frame components may be improved by the provision of strengthening ribs. Figure 2.12 illustrates widely used rib designs for machine columns and their components.

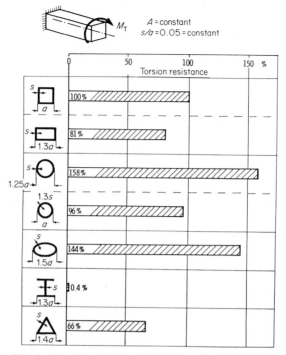

Fig. 2.10 Torsion resistance of differing cross-sections

Examples A to D show longitudinal support ribs, whilst those marked E to H have transverse strengthening ribbing. The relative bending and torsion resistances of these column designs are summarized in Fig. 2.13. It can be seen that when bending loads are applied, longitudinal ribbing improves the rigidity, whilst the addition of an end plate has only a minimal effect. The

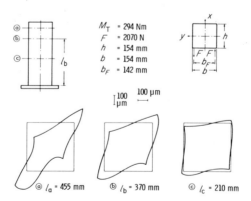

Fig. 2.11 Torsion deformation of an un-ribbed machine column without an end plate

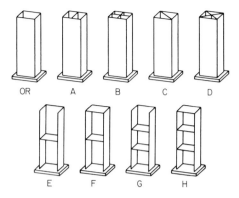

Fig. 2.12 Different rib designs for machine columns

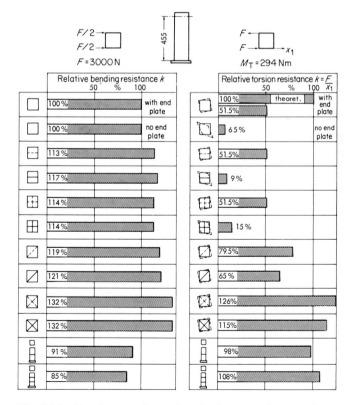

Fig. 2.13 Bending and torsion resistances for machine columns with differing rib designs

reduced stiffness of columns with transverse rib plates, in the models shown here, is due to production faults (welding-zone disturbances).

In the case of torque loading, all forms of ribbing which reduce the degree of deformation of the cross-section improve stiffness. Apart from end plates, transverse and diagonal longitudinal ribs come into this category. For machine columns which are subjected to bending and torque loading simultaneously, double-diagonal longitudinal ribbing with an end plate gives the best result.

If a value analysis is applied to various forms of ribbing, then, apart from stiffness, material volume and the lengths of the weld runs at the joints must also be considered.[4] In contrast to the considerations given above, which are limited to measured deflection criteria, the following examples are the result of a study utilizing the 'finite element' method (see section 2.6).

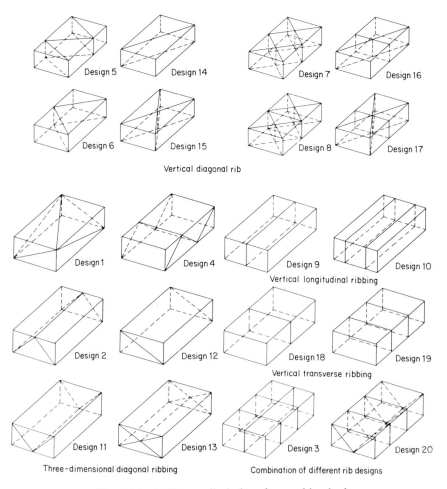

Fig. 2.14 Different rib designs for machine beds

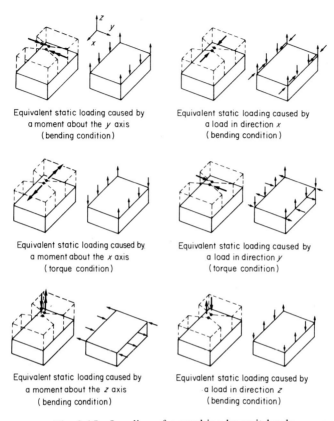

Fig. 2.15   Loading of a machine by unit loads

A model of a machine bed is taken as an example. Figure 2.14 shows the possible variations in ribbing design. The six differing loading conditions depicted in Fig. 2.15 are considered as unit loads which are acting on the model as unit forces in the directions of the co-ordinate axes and as moments about these axes. In order to deduce the total flexibility of a complete machine bed, the flexibility under each of the six loading conditions is summed by arithmetically adding the work done in the deformation for each individual loading condition.

The influence of various ribbing designs on the flexibility material volumes used and lengths of weld runs at the joints of a selection of closed machine-bed designs is summarized in Fig. 2.16. In the left-hand column the flexibility is indicated as percentages of the flexibility of an unribbed design which is used as a basis.

The relatively small difference between the flexibility of the unribbed basic design 0 (100%) and that of the double-diagonal ribbed bed 13 (64%) indicates that the somewhat costly ribbings give little improvement in the flexibility of a frame component. The centre column of the diagram shows that the

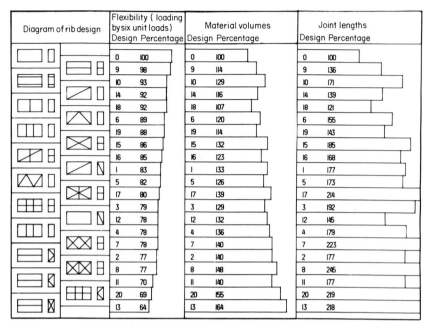

Fig. 2.16 Flexibility, material volumes and joint lengths of closed machine-bed designs

material volume of the differing designs is roughly inversely proportional to their flexibility. The largest material volume is 1.6 times greater than that of the basic design. The various ribbing designs lengthen the joint lengths up to a maximum of 2.5 times that of the basic model, and no direct relationship between these increases in joint length and flexibility can be established.

In order to establish the economic viability of the differing ribbing designs, the product of material volume and flexibility is calculated to evaluate the material utilization; similarly, the welding costs are evaluated by multiplying the joint lengths by the flexibility. These products are illustrated in Fig. 2.17 as histograms for the various ribbings under consideration; from these the most advantageous design may be deduced, i.e. the lowest flexibility for the least material utilization or welding costs. The most advantageous designs appear to be the unribbed basic design, the simple transverse arrangements and the Vee-formed longitudinal ribbings, where the ribs support the guideways. The minimal effect of the vertical longitudinal ribbings in designs 9 and 10 should also be noted.

A frequently used design for machine beds and columns is the box-type design open on one side, or columns without base plates or end plates. These offer some production advantages, whether they be cast or of a fabricated construction. However, when compared with closed-construction forms, designs without a base exhibit a marked increase in torsional flexibility. Fig-

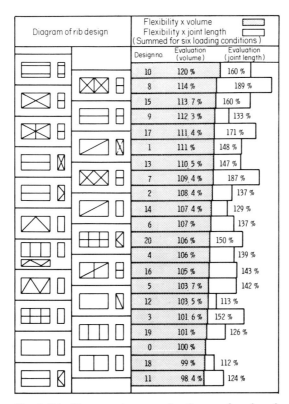

Fig. 2.17 Comparative evaluation of closed machine-frame components

ure 2.18 shows comparative values for fourteen different designs under a torque loading, based on an unribbed closed design. The first six designs, which are only provided with longitudinal or transverse ribs, indicate flexibility up to a hundred times greater than some of the best designs. A real improvement is only apparent when diagonal ribs are introduced, but even here the heavily ribbed designs 7 and 8 still exhibit a 1.5-fold increase in flexibility when compared with closed unribbed designs.

The results of this systematic investigation are contained in a design index.[4] The designer may find therein optimum solutions for typical frame-construction problems. Furthermore, with the help of some rules of application provided, the stiffness may be derived under different loading conditions without the need to perform complicated calculations.[5]

2.3.4.4 Apertures

As may be noted from the preceding sections, the aim should always be to use closed cross-section designs to obtain good bending and torsion stiffness in a machine frame. However, in order to facilitate production and assembly (e.g.

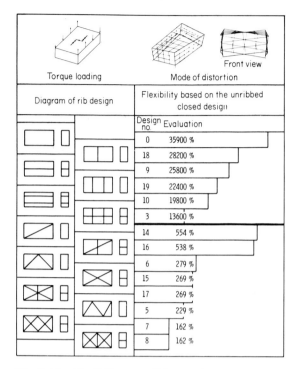

Fig. 2.18 Flexibility conditions of open machine beds under torque loads

extraction of casting sand, mounting of various machine units) it is often necessary to provide apertures in machine-frame components.

Figure 2.19 illustrates the effect of holes on the stiffness of a box girder. The principal result of such apertures is to markedly reduce the torsion stiffness.[6] In order to minimize the effect on the bending resistance, apertures should be as near as possible to the neutral axis.

2.3.4.5 Force transmission

The forces and moments acting on the machine frame are transmitted by slides and guideways into the main body. This causes local distortions, which may have a considerable influence on the total deformation. Typical arrangements of slideways and their supports on the column may be seen in Fig. 2.20. The cross-section designs B, C and G are undesirable, as the elastic flexibility in the walls seriously reduces their stiffness. In the designs marked, A, D, E, F, H and I, the guideways are supported by the side walls. The internal ribbing provided in E, F and I serve on the one hand as very good additional supports for the guideways and on the other to increase the bending and torque resistance of the column.

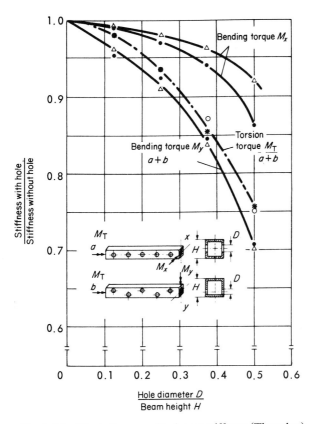

Fig. 2.19 The influence of holes on stiffness (Thornley)

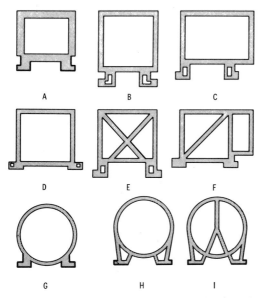

Fig. 2.20 Cross-section designs of machine tool columns

2.3.4.6 Joining methods

Machine-frame components are connected to each other or the machine base by friction or positive clamping devices. The resultant joints influence overall stiffness of the frame as they are usually within the force-flux flow.[7] The various consequential effects of the flexibility at the joints on the displacement of a machine column is diagrammatically illustrated in Fig. 2.21.

A force F_y acting perpendicular to the plane of the diagram would add an additional tangential stress and cause a torque to be acting on the column. Most importantly, any interfaces which are an appreciable distance from the point of loading must have a good stiffness incorporated in the joining method applied, as these have considerable consequential effects due to their unfavourable leverage. The interfaces between various frame components are frequently joined with a number of bolts holding the plane faces together; the rigidity is governed by the following parameters:

(a) shape of the joint face (flange);
(b) bolt patterns and their number;
(c) stiffness of the individual bolt connections.

Joint and flange designs. The force flux in a joint tends to concentrate on those points where the components being held together are pressed against each other by the bolts.

The diversion and concentration of the force flux into the contact areas which are under the influence of the bolts cause a bending stress in the joint and considerable local distortions. Figure 2.22 shows the mechanism of a bending stress in a joint. By positioning the bolts as close as possible to the column wall, such bending stresses may be minimized. Strengthening ribs which support the column wall and the joint in the bolt area improve the

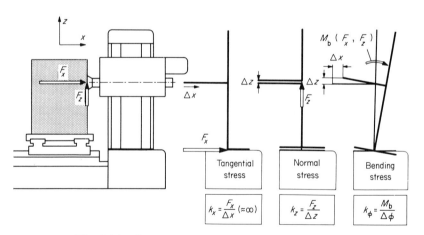

Fig. 2.21 Consequential effects of flexibility at joints

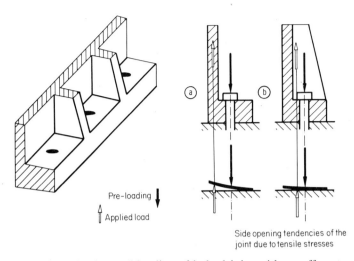

Fig. 2.22 Design and loading of bolted joint with an off-centre bolt pattern

stiffness. An increase in flange thickness has a similar effect. If the bolt pattern is arranged within the plane of the wall, the bending stresses at the joints are obviated. However, the necessary clearances in the wall weaken the cross-section of the frame in the joint area. As the centralized bolt connection, shown on the right of Fig. 2.23, gives considerably better stiffness, the increased production costs involved can in certain cases be justified.

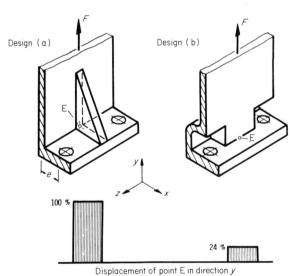

Fig. 2.23 Stiffness comparisons of different joint designs

Number and pattern of bolts in a joint. Wide-reaching research into bolted joints has shown[7] that only a very limited area of a joint around a bolt is active in force transmission. The contact area around the bolt increases with the thickness of the flange due to the effect of a pressure cone, as shown in Fig. 2.24. For maximization of the joint rigidity, the active contact areas should overlap. Hence it is more advantageous, as may be seen from Fig. 2.25, to use a larger number of small diameter bolts closely spaced than a smaller number more widely spaced. Fig. 2.25 shows that with constant load, plate thickness and cross-sectional area of the bolts, the distortions decrease as the number of bolts increase.

Stiffness of the individual bolt connection. The stiffness of an individual bolt connection is dependent upon the stiffness of the bolt itself, as well as that of the flange and the contact area. Figure 2.26 shows a spring analogy of a bolted joint and its components. It may be noted that the joint consists of two spring systems connected in parallel. This analogy is only valid as long as the applied force does not exceed the pre-load of the joint. If this is not so, then in the case of compressive loading only the flange and contact zone are involved in the force transmission, and for tensile loads only the flange and bolt participate. Consequently, correspondingly reduced stiffness results. Thus the bolts should always be pre-loaded with a greater force than the expected applied loading.

A most important component of a bolted joint, as indeed in every other

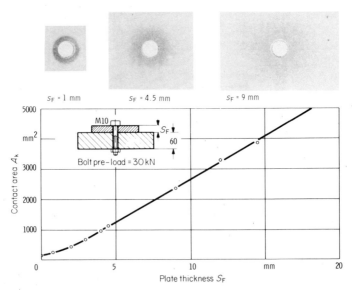

Fig. 2.24 Relationship between effective contact area and joint plate thickness

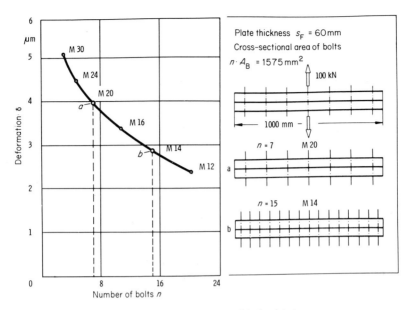

Fig. 2.25 Deformation of bolted joints

form of joint (shrink assemblies, shaft-collar joints, etc.), is the stiffness of the contact zone of the assembled components. The contact area stiffness is largely dependent upon surface conditions in the joint. The surface finish of the faces in contact with each other influences the stiffness conditions. Moreover, the stiffness is improved as the contact pressure is increased because there is an increase in the number of surface peaks touching each

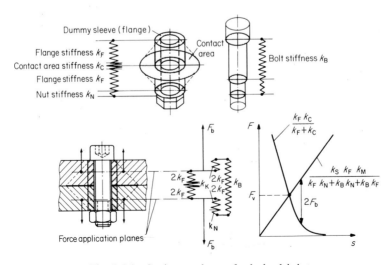

Fig. 2.26 Spring analogy of a bolted joint

other; hence a larger proportion of the area is involved in the force transmissions. The relationship between force and deformation in a contact zone has been established by Levina with an empirical equation. The deformation δ normal to the direction of a mean face pressure is given by

$$\delta = \alpha p^m \qquad (2.6)$$

where δ is in μm and p in Nm$^{-2} \times 10^{-3}$; α and m are dependent upon the surface finish and the materials in contact. For example, when two ground-steel surfaces are in contact, $\alpha = 0.6$ and $m = 0.4$. In general, a good surface finish will give a good contact rigidity.

2.3.5 Design examples

The structural design of machine-frame components is fundamentally dependent upon the duties which they are expected to perform. A few examples are discussed here which are intended to illustrate clearly how closely related the design is to the functional requirements.

Figure 2.27 shows a column of a drilling and milling centre. The basic form of this machine column is a rectangular cross-section. The inside of the column must be kept free to accommodate a balance weight which is required to counteract the weight of the machining head and slides. To improve the torque resistance, incomplete cross ribs are provided, and, in addition, longitudinal ribs give an adequate bending rigidity. The slides are situated too far

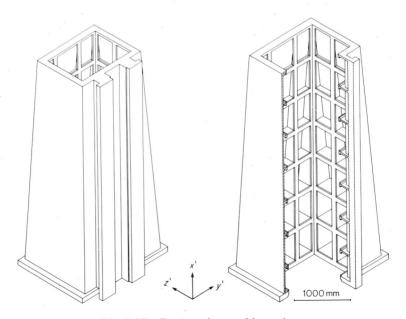

Fig. 2.27   Rectangular machine column

from the side walls and hence are not in the most advantageous position; they are, however, sufficiently supported by the cross ribs.

On the circular column illustrated in Fig. 2.28, the cross-section shows that the slideways have tangential support from the walls of the casting and radial support from the internal ribs. The circular cross-section gives the column very good torque resistance, while the longitudinal ribbing improves the bending resistance about the x and y axes.

The geared headstock illustrated in Fig. 2.29 has internal walls which have other functions besides those of strengthening. Primarily, they are required as spindle supports and to enable the levers to be mounted. Further, these walls have a number of holes bored into them for functional requirements, which consequently reduce the stiffness. Beads and lugs placed around the bores, as well as wall and other internal ribs, serve as stiffeners.

The design of such complicated components with regard to their stiffness frequently results from arbitrary decisions based purely on experience, not knowing whether the stiffness is adequate or even if there is an overprovision, which is of course very costly.

Calculations of static behaviour are possible on modern computers using the 'finite element' method (see section 2.6). The costs involved, however, usually cannot be justified.

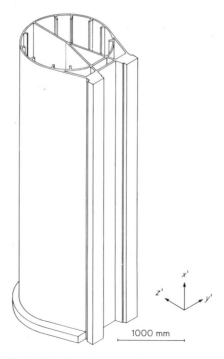

Fig. 2.28 Circular machine column (Heller)

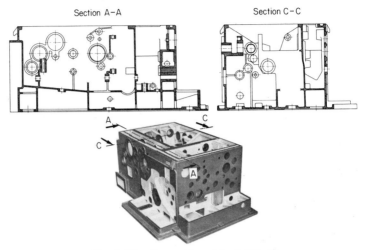

Fig. 2.29 Geared headstock (Cast)

2.4 Design and shape criteria for dynamic loading

2.4.1 Dynamic loads

Apart from the static loads, machine tools are subjected to constantly changing dynamic loading which must be taken into account. Due to the dynamic excitation forces, the whole machine-tool system is subject to vibration. The origin of such vibrations may be classified into external and self excitations, as shown in Fig. 2.30 (see also Volume 4).

In the case of independent excitations, two different types may be iden-

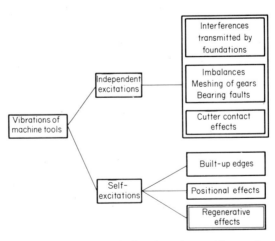

Fig. 2.30 Origins of vibrations in machine tools

tified, i.e. harmonic or continuous excitations on the one hand and pulsating or intermittent excitations on the other.

Typical origins of harmonic excitations are:

(a) unbalanced rotating masses;
(b) bearing irregularities.

Among pulsating excitations, the main causes are:

(a) cutting and forming forces on presses and hammers;
(b) interrupted cuts on metal-cutting machines:
(c) cutter-contact forces when milling;
(d) gear engagement on gear drives;
(e) vibrations transmitted to the machine through its foundation.

Under pulsating excitations, the machine-tool system will mainly vibrate at a natural frequency. In the case of harmonic or continuous excitations, the frequency of the vibration will tend to be that of the exciter frequency. Hence, whenever the frequency of excitation is in the same range as a natural frequency of the machine, particularly large vibration effects are likely to result. This is also the case when the repetition rate of a pulsating excitation coincides with a natural frequency (e.g. an inserted tooth-cutter head). As an example of vibrational deflections, Fig. 2.31 shows the mode shape of the dominant natural frequency of a roll-turning lathe. The applied force in this case acts along the x axis between the tool holder and the workpiece.

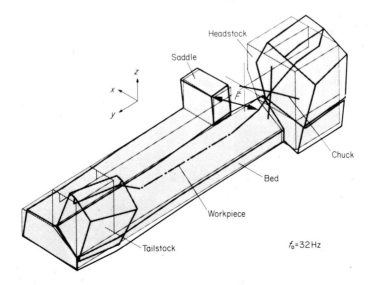

Fig. 2.31 Mode shape of the natural vibrations of a roll-turning lathe

In the case of self excitations, the machine system mainly vibrates at its natural frequency if no external disturbing forces are acting upon it. The vibrations are maintained by the working process itself. The regenerative chatter experienced in almost all metal-cutting operations is typical of such a condition, and is likely to limit severely the performance capabilities of the machine. (The complex relationships of these vibration phenomena are dealt with in detail in Volume 4.)

As the dynamic loads, which are governed by the process, are unavoidable, the vibration amplitudes can only be kept within acceptable limits by an adequate dynamically rigid design.

2.4.2 Dynamic quantifying factors

The dynamic behaviour of a machine-tool frame is dependent upon the following characteristics:

(a) mass and mass distribution;
(b) stiffness;
(c) damping, which is largely governed by the conditions at the interfaces.

Specific vibrational deflections will develop at certain natural frequencies for every system; these are dependent on the above criteria, but are independent of the type of loading. To enable discussion of the dynamic behaviour of such complex structures as machine tools, a knowledge of these vibrational deflections is of prime importance. From this, it is possible to identify the individual machine components from which most of the vibrations originate (analysis of weak points). Figure 2.32 shows the vibrational deflections of a double-gantry milling machine at a natural frequency of 28 Hz. It may be seen that the columns tend to distort in the manner of a cantilever, whilst the cross beam undergoes a severe bending deflection.

Because the system is loaded by a periodic variable force vibrations are induced; the amplitude of the vibrations is not only dependent upon the magnitude and direction of the applied force but also upon the frequency of the excitation. This relationship may be established through the frequency-dependent flexibility $G(j\omega)$, known as the 'flexibility–frequency characteristic'. This requires that the relative dynamic flexibility between the tool and work be determined in all three co-ordinate axes (Volume 4).

The flexibility–frequency characteristic can today also be calculated (see section 2.6), but the costs involved can only rarely be justified. Moreover, there are considerable uncertainties about the accuracy of any result obtained due to the generally unknown rigidity and damping influences of interfaces (flange connections, guideways, bearings, etc.). The upper curve on the left of Fig. 2.33 shows the measured relative flexibility amplitude $G_{xx(\omega)}$ on the x axis for the double-gantry milling machine indicated in Fig. 2.32. The frequencies at which peaks in amplitude are indicated correspond to the natural frequen-

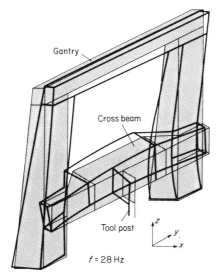

Fig. 2.32 Vibrational deflections of a gantry milling machine

cies of the system. The degree of damping governs the magnitude of the peaks. As already repeatedly mentioned, the relative movements at the interfaces are of major importance here. It may be noted from Fig. 2.32 that in this example the positioning of the guideways on the columns for the cross beam is decisive for the vibrational deformations illustrated. At these points, the vibrational deformation generates a motion perpendicular to the surfaces of the guideways, which have a damping effect due to the squeeze-film effect (see section 4.1).

Apart from the degree of flexibility, the phase relationship between the force being applied and the resultant displacement is also of interest. The lower curve on the left of Fig. 2.33 illustrates this phase relationship. The polar frequency-response locus shown on the right of the diagram gives an equivalent presentation of the amplitude and phase curves combined. For a full analysis of the dynamic behaviour of a machine tool at a given position of its main units, polar frequency response curves are required for the flexibility in all three co-ordinate directions (Volume 4).

2.4.3 *Structural design considerations*

In order to modify and improve the dynamic characteristics, the whole machine system must be considered at the same time, in the same way as described earlier for static considerations. The dynamic characteristics of a machine are primarily dependent upon the static stiffness, as well as the distribution and magnitude of the masses of the individual constructional

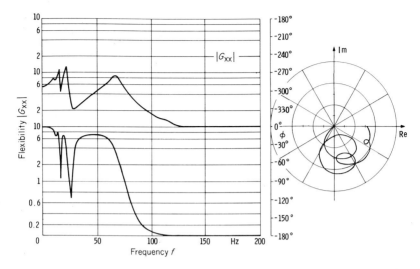

Fig. 2.33 Analysis of the dynamic characteristics of a machine tool using resilience amplitude and phase relationship curves, and a polar frequency response locus

units and the damping effects of the system. The aim should be to produce a construction with low mass values, but at the same time of good rigidity. Most importantly, the mass should be low at those points at which it is reasonable to expect large vibrational amplitudes (e.g. the tool-post weight in the centre of the cross beam in Fig. 2.32). These—generally contradictory—requirements often present great difficulties to the designer. Moreover, the damping effect of the system must be maximized by careful utilization of machine elements with high damping properties, e.g. bolted joints, guideways, etc. To this end, the positioning of the interfaces plays a decisive role in the shape of the vibrational deflections. Relative movement at the interfaces improves damping characteristics on the one hand, but reduces the static stiffness on the other; thus optimum conditions must be determined for these opposing criteria when considering the dynamic characteristics of the whole machine system.[8]

A rough approximation of the influence on the dynamic behaviour can be obtained by considering the oscillations of a single mass, where the relationship for the maximum amplitude at the natural frequency is given by:

$$x_{\text{dyn}_{\text{max}}} = \frac{\tilde{F}}{k} \frac{1}{2D} \tag{2.7}$$

where

$$D = \frac{c}{2m\omega_0} = \frac{X_{\omega=\omega_0}}{2x_{\omega=0}} \tag{2.8}$$

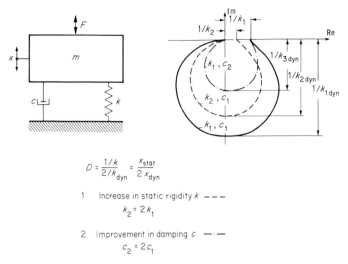

$$D = \frac{1/k}{2/k_{dyn}} = \frac{x_{stat}}{2\,x_{dyn}}$$

1. Increase in static rigidity k – – –
   $k_2 = 2k_1$

2. Improvement in damping c — · —
   $c_2 = 2c_1$

Fig. 2.34 Changes in dynamic behaviour

and

$$\omega_0 = \sqrt{\left(\frac{k}{m}\right)} \qquad (2.9)$$

Therefore:

$$x_{dyn_{max}} = \tilde{F}\frac{1}{c\omega_0} = \tilde{F}\frac{1}{c}\sqrt{\left(\frac{m}{k}\right)} \qquad (2.10)$$

Figure 2.34 shows the effect on a polar frequency response curve from a change in rigidity and damping.

2.4.3.1 Damping in frames

The damping effects within a single-frame component is governed by the mass distribution of the material and its damping properties. Whilst the damping properties of cast iron are higher than for steel, the damping effect at the welded joints of fabricated constructions generally compensates for this disadvantage. To improve the frictional effects within the frame components, the core sand is often left in cast-frame components (dry friction). However, in some cases, a reduction in the damping properties of the system can be detected due to an increase of the mass by the sand.

When considering the whole machine frame, the damping at the interfaces is of great importance. Their influence may be between one or two powers of ten over the damping effect of the material itself. The work done in damping between the surfaces at an interface is converted into heat, and can be traced to minute movements resulting from the displacement of lubricants, and from frictional and deformation forces.[9] The damping effect is dependent upon the magnitude and direction (i.e. tangential or normal to the joint) of these movements. The loading at the joint, which initiates the movement, is governed by the pattern of vibrational distortions. The following parameters influence the damping effect at joints:

(a) geometric shape of joint;
(b) surface finish condition, secondary texture;
(c) contact conditions;
(d) magnitude of contact pressure;
(e) media between the joint faces.

The results of research into the relationship between the damping effect D on the magnitude of the contact pressure on a test column is shown in Fig. 2.35. Observations were carried out at the foot of the column, which was bolted to a base plate. Various surface finishes were assembled with differing contact conditions. The complex mechanism of damping at joints is not yet fully understood. It is extremely difficult to monitor all the many factors which contribute to the total behaviour in a joint. Consequently, it follows that there are no mathematical relationships for dynamic behaviour available which may be applied to obtain precise values for the damping effect. Positive steps which may be taken to improve the damping properties of a system are, for example, the provision of friction plates and auxiliary vibration absorbers,

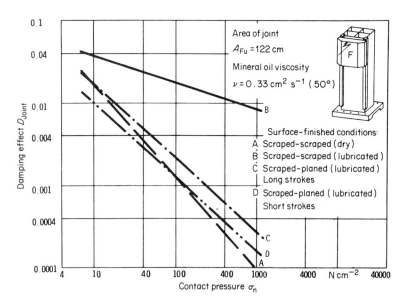

Fig. 2.35 Influence of the surface finish and contact pressure at a joint on the damping effect

which are positioned at points where large vibrational amplitudes may be expected.

2.4.3.2 Auxiliary vibration absorbers, friction dampers

Whenever it is desirable to reduce the vibration amplitude of a natural mode without modifying the frame-component design, then auxiliary vibration absorbers or friction dampers may be used.

To explain the application of auxiliary vibration absorbers, let us consider, for simplicity, the whole system as a single mass of magnitude M vibrating at a natural frequency Ω_0 (see Fig. 2.36). The magnitude of the auxiliary mass M which will be required is dependent upon the permissible resonance magnification, which will result owing to the additional damping effect, i.e.

$$v_{max} = \frac{x_{dynmax}}{x_{stat}} = \sqrt{\left(1 + \frac{2}{\mu}\right)} \quad \text{(when } c = 0\text{)} \qquad (2.11)$$

Given the value of the mass ratio $\mu = m/M$ and the equivalent mass of the system M, which may be established by determining the static stiffness K and the natural frequency $\Omega_0 (M = K/\Omega_0^2)$, the magnitude of the auxiliary mass m required may be calculated from:

$$m = \frac{2M}{v_{max}^2 - 1} \qquad (2.12)$$

The spring rigidity required for the auxiliary vibration damper k and the damping effect c may be calculated as follows:

$$k = m \frac{\Omega_0^2}{(1 + \mu)^2} \Rightarrow \omega_0 = \frac{\Omega_0}{1 + \mu} \qquad (2.13)$$

$$c = 2m\Omega_0 \sqrt{\left[\frac{3\mu}{8(1 + \mu)^3}\right]} \qquad (2.14)$$

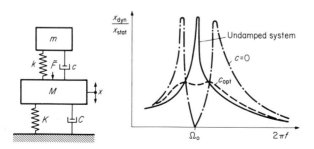

Fig. 2.36 Effect of an auxiliary vibration absorber

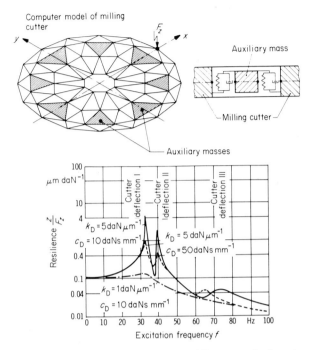

Fig. 2.37 Effects of various stiffness and damping characteristics of auxiliary vibration absorbers (Waldrich Coburg)

As the spring and damping characteristics of the rubber units frequently used are not precisely known, it is necessary to make some adjustments, even when the theoretical optimum conditions have been calculated.

As an example, the calculations for a milling cutter of 5.30 m diameter is illustrated, on which the vibrational distortions are damped with the use of auxiliary vibration absorbers.

Eight auxiliary masses with springs and dampers are fitted on to the computer model shown on the upper left of Fig. 2.37. In the lower part of the diagram, the theoretical flexibility frequencies for the cutter with the auxiliary vibration absorbers are shown for different stiffness and damping characteristics of the auxiliary mass systems. It can be seen that the increasing amplitude for the frequency range under consideration is appreciably reduced only when optimum conditions appertain.

Friction dampers utilize the friction force resulting from the relative movement of the vibrating machine component and a fixed auxiliary mass attached to the machine. In the example of the friction damper in contact with an auxiliary mass depicted in Fig. 2.38, the layout is such that the frictional work done, which is available as damping work, is at a maximum. If the frame component vibrates in direction x at a frequency Ω_0 and with an amplitude x_0,

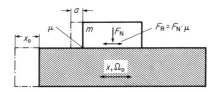

Fig. 2.38 Friction damper

then the auxiliary mass m will rub against the frame with an amplitude a. In the case of dry friction, the frictional force is the product of the normal force F_N and the coefficient of friction μ. The frictional energy (i.e. damping energy) is at an optimum value when the following relationship is satisfied:

$$\frac{F_R}{F_T} = \frac{F_N \mu}{ma\Omega^2} = 0.45 \qquad (2.15)$$

where F_R = frictional force
and F_T = inertia force

The maximum damping energy per period which is then available is given by:

$$E_D = 1.27\, x_0 m \Omega_0^2 \qquad (2.16)$$

where x_0 is the amplitude of the damped system.

2.5 Thermal loading criteria

2.5.1 Thermal loading

In a machine tool, there are a number of heat sources present which cause changes in the temperature distribution within the components, dependent upon the loading conditions and time. Figure 2.39 illustrates the actual and theoretical isothermal lines in the walls of a headstock for an inclined bed lathe, after the spindle has been running for 67 min. The heat sources which affect a machine tool may be classified into two groups:

External heat sources:
(a) temperature of surrounding objects (e.g. heating units, factory walls, other machines) which are at different temperatures from that of the machine;
(b) sun rays;
(c) temperature of ambient media (e.g. air, cooling fluids, lubricants) which change the machine temperature.
In these cases, the heat transfer is basically by radiation and convection.

Internal heat sources:
(a) transmission from electric motors;

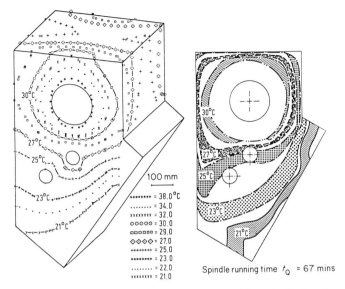

Fig. 2.39 Actual and theoretical isothermal lines in the head stock of an inclined bed lathe

(b) friction in drives and gear boxes;
(c) friction in bearings and guideways;
(d) machining process (cutting action, chips, workpiece).

The effect of the heat sources on the machining process can be presented in the form of a thermal efficiency chain, as shown in Fig. 2.40.[10]

The heat sources cause temperature patterns to be formed in the machine components; their physical magnitude and time of effectiveness are dependent upon their own size and shape, heat conductivity and heat-storage capacity, as well as upon the heat transfer conditions of the surroundings or of other components with which they are in contact.

The thermal deformations which a component undergoes are dependent upon the temperature pattern, its geometry and the fixing or clamping conditions. The distortions between the tool and work at the cutting point are the sum of the deformations of all the components in the chain of deformations. If this results in a relative displacement in one or more of the directions of the tool feeds, then this fault will be detected in its full magnitude as an error in the geometry of the workpiece.

The thermal efficiency chain has a completely different effect in some respects, from the deformations which lie in the force-flux flow due to static and dynamic loads. In contrast to the forces which are acting at the machining point, heat sources can occur at a wide range of different points on a machine tool. Hence, it follows that, under load, all the many individual deformations caused by the process on all the components lying in the force-flux flow contribute to the total deformation at the working point. On the other hand,

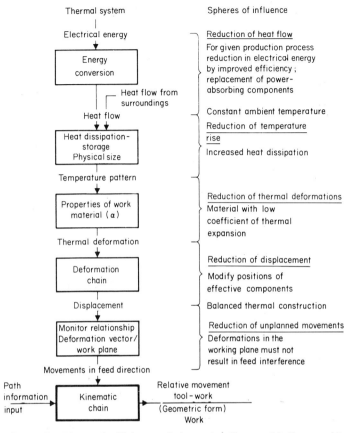

Fig. 2.40 Thermal efficiency chain and spheres of influence (de Haas)

thermal deformations can have a compensating effect. By paying careful attention to the heat-source locations, this counteracting feature can be used to advantage.

Whilst the distortions caused by static and dynamic loads quickly disappear when these loads are removed, the deformations due to the thermal influences gradually increase as the temperature patterns develop, and are still present after the heat source has ceased to be active and until the cooling phase is completed (large time constants). In order to operate production and in particular precision machine tools in near-constant thermal conditions, they are run for some hours before the start of the shift or they are kept at near constant temperatures with the aid of heating or cooling devices.

2.5.2 *Thermal quantifying factors*

The thermal behaviour of a machine tool may be indicated by the relative displacements between the tool and work at the cutting point. Such displacements are governed by the thermal characteristics of all the components which lie in the thermal efficiency chain. The heat sources change the temperature distribution patterns in the components over a period of time, and therefore cause time-dependent distortions.

Hence, the displacements at the cutting point are dependent upon the geometrical construction of the machine and the temperature distribution (θ) at a given time. The latter is governed by the absolute magnitude of the loading (F_0, n_0), the changes in loading ($\Delta F, \Delta n$) and the resultant temperature distribution (θ_0) over the whole machine body.

Generally, the following relationship is valid;

$$\begin{pmatrix} \Delta x \\ \Delta y \\ \Delta z \end{pmatrix} = f((\theta_0), F_0, n_0, \Delta F, \Delta n, t) \qquad (2.17)$$

2.5.3 Structural design considerations

The constructional design opportunities to improve the thermal behaviour of a machine tool fall into two differing categories. On the one hand the aim is to reduce as much as possible all heat sources and thermal influences which are likely to affect the machine. The second approach is to try to ensure that the unavoidable thermoelastic deformations between the tool and work at the cutting point are as small as possible, by considering this factor in the design of the machine components.

The following steps may be considered which will have an influence on the first category of thermal design considerations:

(a) external mounting of drives, i.e. motors and gear boxes;
(b) isolation of the machine from external heat sources (radiation, convection, etc.);
(c) dissipation of frictional heat from bearings and drives;
(d) dissipation of heat generated from the production process by coolant and swarf removal.

As it is not possible to completely eliminate the heat sources, it is necessary to take further steps to minimize the effect on the finished component caused by thermal machine distortions. This task is most complicated because it is necessary to have a knowledge of the isothermal pattern for the whole machine body, and this is constantly changing in relation to varying loading conditions. The options which the designer may consider to this end are:

(a) the positioning of the heat sources in the machine frame;
(b) the distribution of the masses in the frame which will be heated and cooled;
(c) the mounting positions of the machine components, so that the deformations are minimized in the critical directions (see Fig. 2.41);

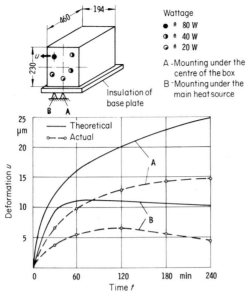

Fig. 2.41 Influence of mounting position on deformation

(d) the design of the heat transfer in a controlled transmission pattern or the blockage of the heat flow;
(e) the incorporation of expansion joints without a material deterioration of the static and dynamic characteristics.

An accurate calculation of the thermoelastic behaviour of a complex machine structure is only possible with the aid of a digital computer (see section 2.6).

As an example, Fig. 2.42 illustrates a headstock of an inclined bed lathe, where the instantaneous displacement of the centre of the spindle is considerably reduced in the direction of the tool feed (direction v), compared with the vertical deformation u. The distortion v introduces an error directly on the diameter being turned, whereas a movement in direction u has only a minor effect on the production result.

Figure 2.43 illustrates a machine-tool column, with a flange-mounted motor of about 3 kW head load fitted to the rear wall. Due to the one-sided heating of the column, a distortion is produced which is caused by the greater expansion of the rear wall of the column. This effect may be reduced by a reduction in the thickness of the rear wall and the introduction of expansion joints. Care must be taken in such instances to ensure that the static and dynamic stiffness of the construction are not seriously reduced.

Further opportunities to reduce the effect of thermoelastic deformations exist through the application of feedback controls and the utilization of compensating devices in the control system (geometric adaptive control) (see

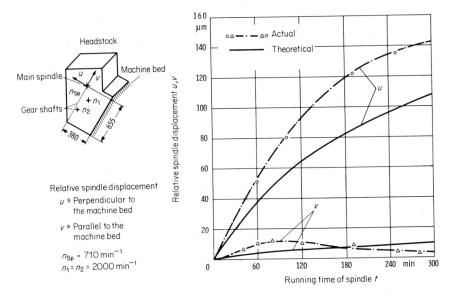

Fig. 2.42 Spindle displacements on an inclined bed lathe

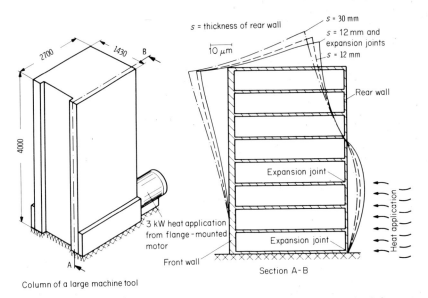

Fig. 2.43 Influence of constructional changes on the thermoelastic deformations of a machine tool

Volume 3). However, these methods will invariably lead to a greater overall machine cost.

2.6 Mathematical analysis of machine-tool components

2.6.1 Fundamentals

A formulation of the physical/mathematical interrelationships is necessary for a clear understanding of the static, dynamic and thermal characteristics of machine-tool-frame components. In simple cases the basic equations for elastic deformations may be applied and by considering the given boundary conditions the problems can be completely solved.

When an exact solution is not possible, due to the complexity of the component, then 'approximation methods' are available, as shown in Fig. 2.44. In such cases one may differentiate between results which, as in the case of exact solutions, portray the behaviour of the structures in a continuous mode or at predetermined discrete points. In the former case, functional analysis methods are used to give a 'closed approximation', e.g. the 'Ritz' method; the latter produces 'discrete approximations', the two best-known methods for finding them being the 'difference equations' and the 'finite element' techniques.

In the case of 'difference equations', the differential coefficients (or quotients) which occur in the equations and boundary conditions for a structure are replaced by difference quotients. The resulting system of algebraic equations are then solved by well-known mathematical methods.

The 'finite element' method is based on the concept of subdividing a given structural shape into simply defined elements with clear boundaries. Approximate relationships are formulated for each of these elements, which consider the required variables of state, e.g. displacements, stress and temperature.

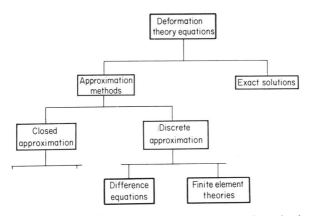

Fig. 2.44 Possible solutions of equations for elastic deformations theory

The mathematical complexity of both the difference equations as well as the finite element methods is such that the calculations can only be economically carried out with the aid of a digital computer.

For the mathematical analysis of machine frames and their components, the finite element method is the most widely used technique.[11 to 13]

2.6.2 Fundamentals of the finite element method

The basic procedure for applying the finite element method is diagrammatically presented in Fig. 2.45. The geometry of the structure to be analysed is approximated by elements with simple boundaries and easily defined dimensions. This subdivision of the component into a mathematical model must be done in such a way that the choice of the types of elements and their distribution reflects the geometric form and the expected deformations. To facilitate this requirement a large number of element types has been established.[12]

The simplest are truss elements which cater for single-axis stress conditions and rectangular beams which may also incorporate loads perpendicular to the length axis.

For conditions in planes, triangular or rectangular membrane elements may be used. For the approximation of thin-walled structures which are additionally loaded perpendicularly to their main plane, plate-bending and shell elements are available. For complex three-dimensional geometries on which a three-dimensional stress pattern is expected, three-dimensional finite elements are used, e.g. tetrahedral and hexahedral elements. A selection of the most common elements is illustrated in Fig. 2.46.

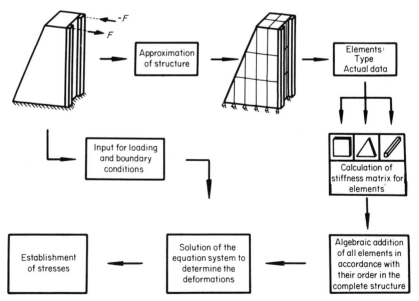

Fig. 2.45 Procedure for the application of the finite element method

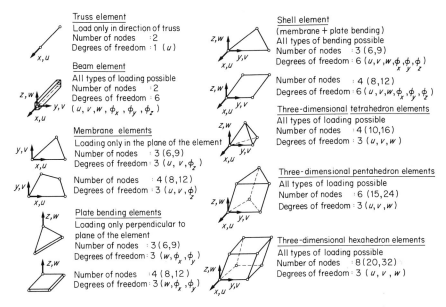

Fig. 2.46 Various types of element for finite element analysis

The appropriate elements are connected at the nodes which are always on the corners and edges to represent the complete system. As the characteristics of the elements are concentrated in the node positions, the characteristics of the complete structure are obtained through these discrete points after the summation of the elements in accordance with their position within the whole component. This discrete mathematical model consists of a group of equations which consider the external loads, the boundary conditions such as those introduced by clamping, as well as the node-connection conditions of the elements:

$$\{F\} = [K]\{U\} \qquad (2.18)$$

In this relationship $\{F\}$ is the vector, whose components encompass all external forces and moments. The matrix $[K]$ gives the stiffness conditions of the structure and the vector $\{U\}$ represents the displacements and distortions at all the discrete node points. The solution of this group of equations—which may have several thousand unknowns in the more complex examples—requires the use of a suitably powerful digital computer.

2.6.2.1 *Derivation of a matrix for the rigidity of an element*
The example of a tension–compression finite element shows the derivation of a matrix for the stiffness of the element based on the displacement equation. The push–pull truss is shown in the upper right of Fig. 2.47. The length of the truss is L, and the co-ordinate x connects nodes 1 and 2.

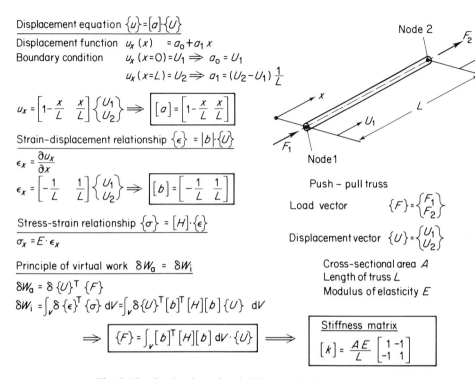

Fig. 2.47 Derivation of a rigidity matrix for a push–pull truss

The deformation behaviour between these nodes can be expressed with a linear polynomial equation:

$$U_x(x) = a_0 + a_1 x \tag{2.19}$$

The two unknowns a_0 and a_1 may be determined from the displacement of the nodes:

$$U_{x(x=0)} = U_1 \Rightarrow a_0 = U_1 \tag{2.20}$$

$$U_{x(x=L)} = U_2 \Rightarrow a_1 = \frac{1}{L}(U_2 - U_1) \tag{2.21}$$

From this displacement at a given position x may be determined, and may be expressed in relation to the displacement of the nodes thus:

$$U_x(x) = \begin{bmatrix} 1 - \dfrac{x}{L} \; ; \; \dfrac{x}{L} \end{bmatrix} \begin{Bmatrix} U_1 \\ U_2 \end{Bmatrix} \tag{2.22}$$

or, in general terms:

$$\{U\} = [a] \begin{Bmatrix} U_1 \\ U_2 \end{Bmatrix} \tag{2.23}$$

The expansion within an element may be determined by the partial differentiation of the displacement relationship. In the case of the tension–compression truss, this is given by:

$$\varepsilon_x = \frac{\partial U_x(x)}{\partial x} = \frac{\partial}{\partial x}\left[1 - \frac{x}{L} ; \frac{x}{L}\right]\begin{Bmatrix} U_1 \\ U_2 \end{Bmatrix} = \frac{\partial [a]}{\partial x}\begin{Bmatrix} U_i \\ U_2 \end{Bmatrix} \quad (2.24)$$

$$\varepsilon_x = \left[-\frac{1}{L} ; \frac{1}{L}\right]\begin{Bmatrix} U_1 \\ U_2 \end{Bmatrix} \quad (2.25)$$

and therefore:

$$\{\varepsilon\} = [b]\{U\}, \quad \text{where } [b] = \frac{\partial [a]}{\partial x} \quad (2.26)$$

The material characteristics are assumed to be linear–elastic and isotropic. The stress–strain relationship is governed by Hooke's law, i.e.:

$$\{\sigma\} = [H]\{\varepsilon\} \quad (2.27)$$

where for a truss:

$$[H] = E$$

The stiffness matrix is derived from the 'principle of virtual work'. This states that the external virtual work δW_a which is obtained from the external load $\{F\}$ with the actual displacement $\delta\{U\}^T$ equals the internal work done δW_i which results from the stress and the actual strain $\delta\{\varepsilon\}^T$, i.e.:

$$\delta W_a = \delta W_i \quad (2.28)$$

$$\delta\{U\}^T\{F\} = \int_v \delta\{\varepsilon\}^T\{\sigma\}\, dV \quad (2.29)$$

Substituting equations (2.26) and (2.27) into the right-hand side of equation (2.29) we get:

$$\delta\{U\}^T F = \int_v \delta\{U\}^T[b]^T[H][b]\{U\}\, dV \quad (2.30)$$

and hence

$$\{F\} = \int_v [b]^T[H][b]\, dV\{U\} \quad (2.31)$$

This integral relates the force vector to the displacement vector, and is the general equation for the determination of the stiffness matrix for given elements:

$$[K] = \int_v [b]^T[H][b]\, dV \quad (2.32)$$

Substituting the values for the tension–compression truss, we get:

$$[K] = \int_{x=0}^{x=L} \begin{bmatrix} -\dfrac{1}{L} \\ \dfrac{1}{L} \end{bmatrix} E \begin{bmatrix} -\dfrac{1}{L}; \dfrac{1}{L} \end{bmatrix} A\,dx \qquad (2.33)$$

$$[K] = \dfrac{AE}{L} \begin{bmatrix} 1 & -1 \\ -1 & 1 \end{bmatrix} \qquad (2.34)$$

As the elements become more and more complex (type of element, number of degrees of freedom, order of the displacement equation, i.e. the number of nodes), so the order of the rigidity matrix rises disproportionally. For example, for a flat rectangular plain-stress element, with two degrees of freedom for each node, which typifies a plane-strain condition, we get the stiffness matrix of the order of 8×8 as shown in Fig. 2.48.

2.6.2.2 Superimposing stiffness matrices of elements to obtain the complete stiffness matrix

By superimposing the stiffness matrices of elements, the finite element approximation method is used to obtain the global stiffness matrix of a complete structure. This is illustrated with a simple and easily followed example.

Figure 2.49 shows a truss structure consisting of three elements e_1, e_2 and e_3 which are connected to each other at node points 2 and 3, and are fully constrained at node points 1 and 4 in the foundation. The geometry of the structure is known and expressed in terms of the lengths L, areas A, modulus of elasticity E and the applied forces. The displacements U_2 and U_3 at node points 2 and 3 are to be determined.

The stiffness matrices of the elements i are found from equation (2.34) as:

$$K_i = \alpha_i \begin{bmatrix} 1 & -1 \\ -1 & 1 \end{bmatrix}$$

where

$$\alpha_i = \dfrac{A_i E_i}{L_i}$$

The following two conditions must be satisfied to enable the distortion characteristics of the complete structure to be determined.

Compatibility conditions. The displacements at the nodes is the same for all the adjoining elements:

$$U_2^{e1} = U_2^{e2} = U_2 \quad \text{and} \quad U_3^{e2} = U_3^{e3} = U_3$$

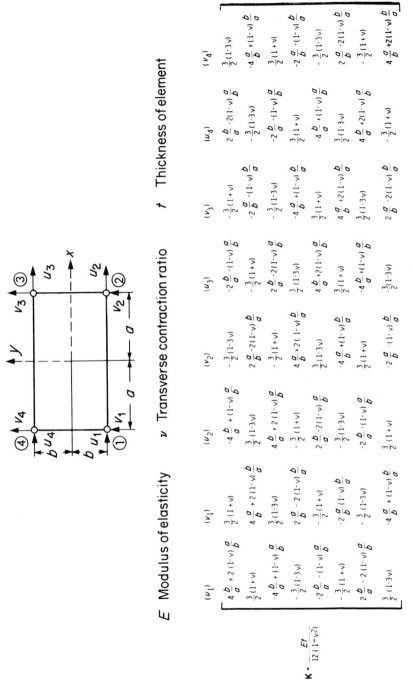

Rectangular plate element

E Modulus of elasticity ν Transverse contraction ratio t Thickness of element

$$K = \frac{Et}{12(1-\nu^2)}$$

$$
\begin{bmatrix}
& (u_1) & (v_1) & (u_2) & (v_2) & (u_3) & (v_3) & (u_4) & (v_4) \\
(u_1) & 4\tfrac{b}{a}+2(1-\nu)\tfrac{a}{b} & \tfrac{3}{2}(1+\nu) & -4\tfrac{b}{a}+(1-\nu)\tfrac{a}{b} & -\tfrac{3}{2}(1-3\nu) & -2\tfrac{b}{a}-(1-\nu)\tfrac{a}{b} & -\tfrac{3}{2}(1+\nu) & 2\tfrac{b}{a}-2(1-\nu)\tfrac{a}{b} & \tfrac{3}{2}(1-3\nu) \\
(v_1) & \tfrac{3}{2}(1+\nu) & 4\tfrac{a}{b}+2(1-\nu)\tfrac{b}{a} & \tfrac{3}{2}(1-3\nu) & 2\tfrac{a}{b}-2(1-\nu)\tfrac{b}{a} & -\tfrac{3}{2}(1+\nu) & -2\tfrac{a}{b}-(1-\nu)\tfrac{b}{a} & -\tfrac{3}{2}(1-3\nu) & -4\tfrac{a}{b}+(1-\nu)\tfrac{b}{a} \\
(u_2) & -4\tfrac{b}{a}+(1-\nu)\tfrac{a}{b} & \tfrac{3}{2}(1-3\nu) & 4\tfrac{b}{a}+2(1-\nu)\tfrac{a}{b} & -\tfrac{3}{2}(1+\nu) & 2\tfrac{b}{a}-2(1-\nu)\tfrac{a}{b} & -\tfrac{3}{2}(1-3\nu) & -2\tfrac{b}{a}-(1-\nu)\tfrac{a}{b} & \tfrac{3}{2}(1+\nu) \\
(v_2) & -\tfrac{3}{2}(1-3\nu) & 2\tfrac{a}{b}-2(1-\nu)\tfrac{b}{a} & -\tfrac{3}{2}(1+\nu) & 4\tfrac{a}{b}+2(1-\nu)\tfrac{b}{a} & \tfrac{3}{2}(1-3\nu) & -4\tfrac{a}{b}+(1-\nu)\tfrac{b}{a} & \tfrac{3}{2}(1+\nu) & -2\tfrac{a}{b}-(1-\nu)\tfrac{b}{a} \\
(u_3) & -2\tfrac{b}{a}-(1-\nu)\tfrac{a}{b} & -\tfrac{3}{2}(1+\nu) & 2\tfrac{b}{a}-2(1-\nu)\tfrac{a}{b} & \tfrac{3}{2}(1-3\nu) & 4\tfrac{b}{a}+2(1-\nu)\tfrac{a}{b} & \tfrac{3}{2}(1+\nu) & -4\tfrac{b}{a}+(1-\nu)\tfrac{a}{b} & -\tfrac{3}{2}(1-3\nu) \\
(v_3) & -\tfrac{3}{2}(1+\nu) & -2\tfrac{a}{b}-(1-\nu)\tfrac{b}{a} & -\tfrac{3}{2}(1-3\nu) & -4\tfrac{a}{b}+(1-\nu)\tfrac{b}{a} & \tfrac{3}{2}(1+\nu) & 4\tfrac{a}{b}+2(1-\nu)\tfrac{b}{a} & \tfrac{3}{2}(1-3\nu) & 2\tfrac{a}{b}-2(1-\nu)\tfrac{b}{a} \\
(u_4) & 2\tfrac{b}{a}-2(1-\nu)\tfrac{a}{b} & -\tfrac{3}{2}(1-3\nu) & -2\tfrac{b}{a}-(1-\nu)\tfrac{a}{b} & \tfrac{3}{2}(1+\nu) & -4\tfrac{b}{a}+(1-\nu)\tfrac{a}{b} & \tfrac{3}{2}(1-3\nu) & 4\tfrac{b}{a}+2(1-\nu)\tfrac{a}{b} & -\tfrac{3}{2}(1+\nu) \\
(v_4) & \tfrac{3}{2}(1-3\nu) & -4\tfrac{a}{b}+(1-\nu)\tfrac{b}{a} & \tfrac{3}{2}(1+\nu) & -2\tfrac{a}{b}-(1-\nu)\tfrac{b}{a} & -\tfrac{3}{2}(1-3\nu) & 2\tfrac{a}{b}-2(1-\nu)\tfrac{b}{a} & -\tfrac{3}{2}(1+\nu) & 4\tfrac{a}{b}+2(1-\nu)\tfrac{b}{a}
\end{bmatrix}
$$

Fig. 2.48 Stiffness matrix for a rectangular bending element.

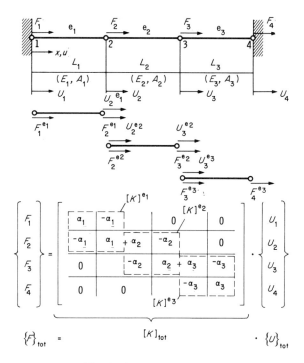

Fig. 2.49 Truss structure

Equilibrium condition. The sum of the outer and inner forces equals zero:

$$F_1 = F_1^{e1}$$
$$F_2 = F_2^{e1} + F_2^{e2}$$
$$F_3 = \phantom{F_2^{e1} +{}} F_3^{e2} + F_3^{e3}$$
$$F_4 = \phantom{F_2^{e1} + F_2^{e2} +{}} F_4^{e3}$$

If the element forces are expressed in terms of the spring constants α and the node displacements U we have:

$$F_1 = \alpha_1(U_1 - U_2)$$
$$F_2 = \alpha_1(-U_1 + U_2) + \alpha_2(U_2 - U_3)$$
$$F_3 = \alpha_2(-U_2 + U_3) + \alpha_3(U_3 - U_4)$$
$$F_4 = \alpha_3(-U_3 + U_4)$$

After multiplying out these expressions, we obtain the following matrix:

$$\begin{Bmatrix} F_1 \\ F_2 \\ F_3 \\ F_4 \end{Bmatrix} = \begin{bmatrix} \alpha_1 & -\alpha_1 & 0 & 0 \\ -\alpha_1 & \alpha_1+\alpha_2 & -\alpha_2 & 0 \\ 0 & -\alpha_2 & \alpha_2+\alpha_3 & -\alpha_3 \\ 0 & 0 & -\alpha_3 & \alpha_3 \end{bmatrix} \begin{Bmatrix} U_1 \\ U_2 \\ U_3 \\ U_4 \end{Bmatrix} \quad (2.35)$$

$$\{F\}_{tot} = [K]_{tot} \{U\}_{tot}$$

(with submatrices $[K]^{e_1}$, $[K]^{e_2}$, $[K]^{e_3}$ indicated)

It can now be clearly seen how the stiffness matrices of individual elements are combined to obtain an overall stiffness matrix in respect of the complete structure.

If the boundary conditions, i.e. the displacements of the node points 1 and 4, are now considered:

$$U_1 = U_4 = 0$$

and at the same time it is assumed that the lengths and the moduli of elasticity of the individual trusses are equal, then the following relationships for the node displacements U_2 and U_3 ensue:

$$U_2 = \frac{2F_2 + F_3}{3\alpha}$$

$$U_3 = \frac{F_2 + 2F_3}{3\alpha}$$

The procedure for establishing a global stiffness matrix and the solution of the system equation may be mechanized by the use of a digital computer in accordance with established practice. The input will consist of the geometric data of the mathematical model (e.g. node co-ordinates, node–element sequences, element thicknesses), the boundary conditions (e.g. fixed points, conditions of symmetry) and the external loadings consisting of applied forces and moments.

The results of the calculations vary depending on the objective and are explained in greater detail in the following sections.

2.6.3 Survey of calculations made possible using the finite element method

The presentation of the fundamental principles for calculating the elastic characteristics of machine components using the finite element method has already shown that it is only practical with the use of electronic data-processing equipment.

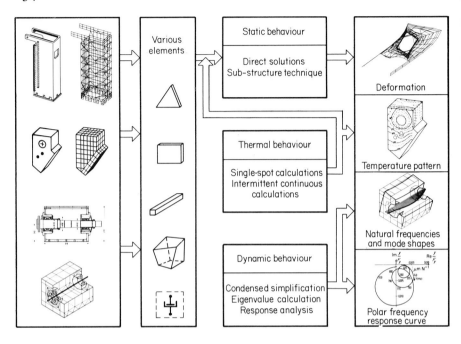

Fig. 2.50 Calculation system for structure analysis using the finite element method

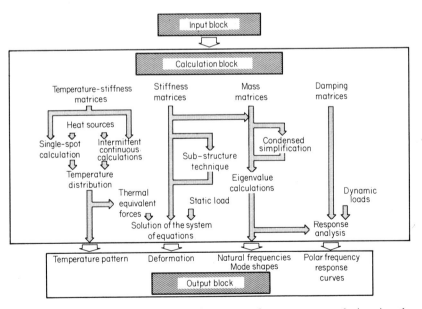

Fig. 2.51 Components of a calculating system for structure analysis using the finite element method based on the computer programs FINEL, FINELT and DYNFIN

Analysis of the static, dynamic and thermal behaviours of machine components requires the availability of a comprehensive range of programs which incorporate an adequate variation of elements and the appropriate calculating algorithm.[12]

Figure 2.50 depicts the operational capacity of a calculation system for structural analysis utilizing the finite element method. In the diagram presented in Fig. 2.51, the components of the system are shown, as well as the options available for using it, governed by the given requirements.

2.6.4 Calculation examples

2.6.4.1 Calculations for the static behaviour of machine-frame components

The bed of an external cylindrical grinding machine, shown in Fig. 2.52, is used as an example for the establishment of the stiffness of a complicated machine component.[13] The sketch on the upper left of the diagram shows the general arrangement of the basic units of the machine in block form. The cellular-constructed bed is heavily ribbed in order to obtain a good stiffness, necessary for the process accuracy demanded from the machine. A fundamental criterion for the quality of the machine output is the relative displacement between the grinding wheel and the work being ground. Hence the aim of the mathematical investigations is to establish the distortion pattern in the plane of the work area. The computer-produced drawing (plotter output) of the mathematical model, containing 192 nodes, each with 6 degrees of freedom, is reproduced on the left of the diagram. It may be noted that the structure of the bed is approximately subdivided into rectangular, triangular and truss elements, to enable the mathematical analysis to be carried out. The reactions to the grinding forces, produced in the plane of the work area, were

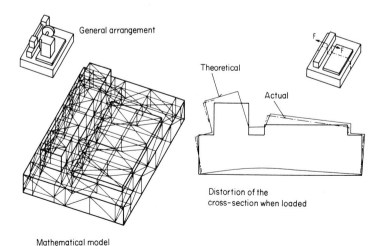

Fig. 2.52 Distortion analysis of a grinding-machine bed

considered as the loading of the structure. In the lower right of Fig. 2.52 the theoretical deformations of the cross-section of the bed in the plane of the work area is shown.

By a careful analysis of such distortion data, the designer is given objective information to introduce extra stiffening sections, thus improving the structure.

2.6.4.2 Calculations for the dynamic behaviour of machine-frame components
When examining the dynamic problems utilizing the finite element method, the mass and damping effects acting in the structure must also be taken into account. Hence, the dynamic behaviour of a structure may be determined by a set of linear differential equations of the second order:

$$[M]\{\ddot{U}\} + [C]\{\dot{U}\} + [K]\{U\} = \{F(t)\} \tag{2.36}$$

For given starting and boundary conditions, the time-dependent values (deformations, velocities and accelerations) may be obtained by the integration of the differential equation system. In addition to the stiffness matrix, a mass matrix and damping matrix must be established for the solution of such dynamic problems.[14] The overall cost of calculations for dynamic behaviour is considerably higher than that of calculations for static behaviour, due to the additional computer time and capacities required. For this reason, the mathematical model must be 'condensed', i.e. the characteristics of the model 'reduced' to:

(a) the nodes and degrees of freedom where forces are acting or a mass is active and
(b) the interfaces of the construction's components

where the results are going to be of some material benefit.

Figure 2.53 shows this procedure using a simple example. In the centre, the complete mathematical model is shown, the condensed version being illustrated on the right.

Firstly, the natural frequencies and the normal natural-mode shapes are established for the model, neglecting any damping effects. This is followed by a calculation for the flexibility–frequency characteristics of the structure, where the damping effects are taken into account.

Whilst the mass and stiffness matrices may generally be constructed with a reasonable degree of accuracy, this is not possible in the case of the damping matrix. As already mentioned in section 2.4.2, the determination of damping values is, at present, still very difficult.

The vertical boring machine shown in Fig. 2.54 is now considered from the point of view of its dynamic characteristics. The mathematical model contains 1093 elements with 617 nodes. The walls of the machine are approximated by plain cup elements, the guides by beam elements and the interfaces of the individual components by special coupling elements (spring and damper).

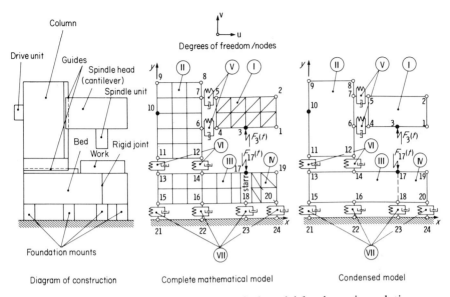

Fig. 2.53 Construction of mathematical model for dynamic analysis

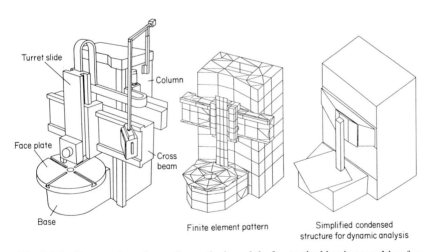

Fig. 2.54 Preparation of a mathematical model of a vertical boring machine for dynamic analysis

The structure was then 'condensed' into the outline, as depicted on the right of the figure.

The results of the calculations produced the natural frequencies and their associated nominal mode shape, two examples of which are shown in Fig. 2.55 just as the plotter of the computer produced them. At the natural frequency of 63 Hz, the turret slide is seen to tend to tilt, whilst at the frequency

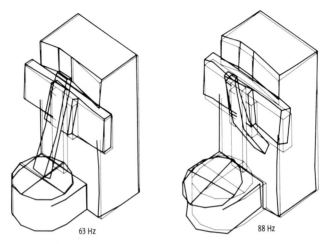

Fig. 2.55 Calculated natural vibration patterns

of 88 Hz considerable slip of the turret slide on the cross beam is observed, as well as a bending tendency of the column.

The relative flexibility–frequency response in direction x, $G_{xx}(f)$, between the face plate and the turret slide was calculated for a frequency range of 0 to 250 Hz, and consisted of two absolute frequency characteristics (between the underside of the ram and the point on the face plate below). Figure 2.56 shows one of the absolute frequency-response curves. The dominant resonance frequencies of 63 and 88 Hz correspond to the natural-mode shapes depicted in Fig. 2.55.

Owing to the high costs involved in the computations and the inaccuracies expected in the results due to the mostly incomplete knowledge of the damp-

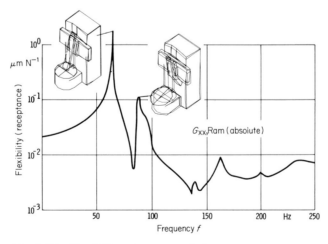

Fig. 2.56 Flexibility (receptance)–frequency response curve of the underside of the ram (tool)

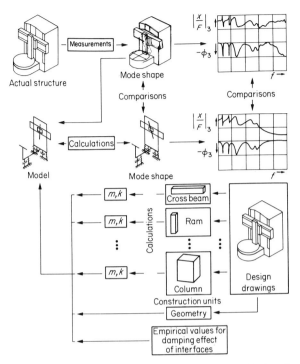

Fig. 2.57 Procedure for the combined theoretical and measured distortion analysis to improve the dynamic behaviour

ing values, a purely theoretical analysis for complex structures is only applied in urgent and exceptional circumstances. Once the machine has been constructed, a practical measured analysis is more economical and more informative (see Volume 4).

The analytical studies for improving the construction design of machines currently utilize both approaches. A simplified mathematical model is constructed based on the information contained in the design drawings. This simulation, which basically incorporates the most important variable elements of the structure, is continuously modified until its dynamic characteristics (natural frequencies, mode shapes and flexibility values) correspond to the measured values (model analysis); see Fig. 2.57. Individual parameters may then be varied on the mathematical model and the effect on the dynamic characteristics can be studied. When a satisfactory combination of all the parameter values is obtained, they must then be translated into the actual construction.

2.6.4.3 Calculations for the thermal behaviour of machine-frame components
The calculations for the thermal behaviour of machine bodies may be broadly classified into two groups, as indicated in Fig. 2.51. Firstly, the instantaneous temperature distribution in the complete component is calculated. The results

of such calculations for the thermal effects are used as input values for the actual deformation calculation process.[15]

To establish the time-dependent temperature pattern a finite element mathematical model is constructed, which permits the calculations to be carried out based on the heat sources and heat sinks. The time-dependent constant, as well as variable heat sources, must be considered. The isothermal pattern is calculated for given time intervals and graphically presented as shown in Fig. 2.39. In addition to the calculations of the temperature patterns after a given heating time, it is also possible to determine the stationary temperature effects.

The temperature patterns, i.e. the temperatures at the nodes of the mathematical model, are then utilized to proceed with the deformation calculations. The theoretical thermoelastic distortions are translated into static deformation calculations. This enables the thermal-equivalent forces to be formulated, which take into account the linear expansions of the elements resulting from the temperature changes.

The thermal-equivalent force is defined as the force which would cause the same deformation of an element as the temperature change. The thermal-equivalent force for a truss element is shown in Fig. 2.58.

After the values of the thermal-equivalent forces have been established the deformations are added to those calculated for the static behaviour, where the thermal-equivalent forces are considered to be acting as external forces at all nodes in the system.

As an example, the theoretical thermal behaviour of a machine frame is considered. Figure 2.59 shows a sketch of the column on the left and its finite element structure used for the calculations on the right. A time-dependent room-temperature variation of 10 K is assumed to be acting. The results, i.e.

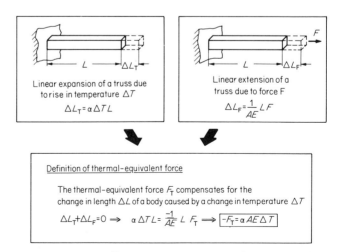

Fig. 2.58 Thermal equivalent force for a truss

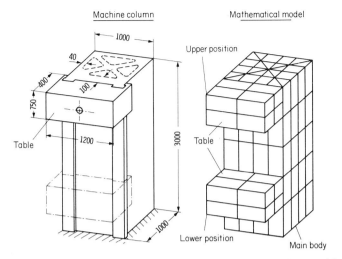

Fig. 2.59 Schematic representation of a machine column with its mathematical model

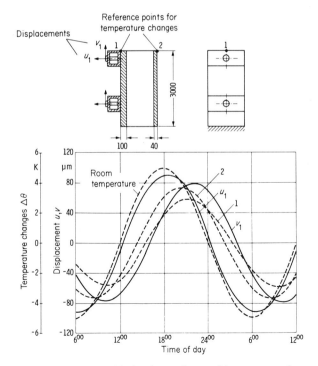

Fig. 2.60 Thermal behaviour of a machine column due to temperature changes

the consequential displacements u and v at two different positions on the column together with the temperature variations at three points, are illustrated in Fig. 2.60. The variations in displacement $(u_1 - u_2)$ and $(v_1 - v_2)$ have in this case the effect of introducing geometric errors into the workpiece.

As in the case of the dynamic behaviour of machine tools, where an inadequate knowledge of the damping effects at the interfaces made an exact solution very difficult, the thermal calculations are hindered because only an approximation of boundary conditions such as heat transmissions, heat-transfer coefficients and radiation effects is known, making a precise prediction of thermoelastic deformation very difficult. However, although the exact values cannot be obtained, the establishment of trends and expected variations are adequate to enable the designer to take the appropriate action.

2.7 Noise reduction in machine design

As a result of the efforts made in recent years to improve the environmental conditions of the machine shop, the noise emission of machine tools has become a major quality criterion (see also Volume 4).[16 to 18] Consequently, it is necessary for the designer to understand the basic concepts of noise generation and noise reduction, so that he may take these into account when considering the design and the application of the machine.

Figure 2.61 illustrates a typical pattern of noise emission from a metal-cutting machine tool in relation to load, rotational speed and torque.[17] As the load increases (torque × speed) the radiated acoustic power generally

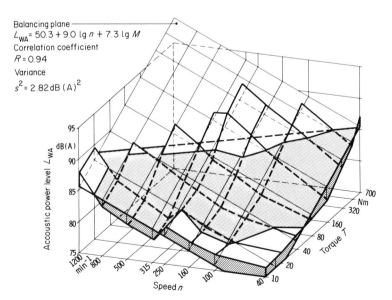

Fig. 2.61 Accoustic power readings in relation to machine loading data

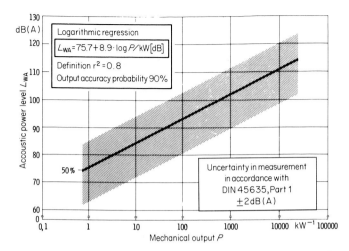

Fig. 2.62 Accoustic power readings on a drive unit at full rated load

increases at the same time. This is also valid for a gear-driven drive (Fig. 2.62).[19] The graph indicates the mean values as well as the scatter band of many readings taken with an acoustic power meter on a working drive unit.

The acoustic power is determined by a standardized practice which aims to ensure that the values can be usefully interpreted. The procedure adopted and the analysis technique used to determine the source of the noise emission are both explained in detail in Volume 4 of this series.

2.7.1 Fundamentals

Noise may be generated in two different ways: by surface vibratory movements of the machine-construction units, tools and workpieces (caused by dynamic excitation forces), on the one hand, and as sound transmitted by air in the form of air currents, flow or turbulence, on the other. These relationships are presented in Fig. 2.63.

Air turbulence caused by fast-rotating three-jaw chucks, suction noise of ventilators and air expelled from pneumatic installations are examples of direct air-borne sounds. In the latter case, silencers or mufflers[20] are frequently suitable devices to use as noise reducers.

Indirect air-borne sound generation due to the mechanical vibrations of machine components are caused by the varying forces of the process and the forces acting internally in the machine. The mechanical constructional components of the machine, the work and the tool are excited into vibrations at the point of force application, with a velocity v_1 due to the dynamic forces (Fig. 2.64).[21]

In accordance with their dynamic transmission characteristics, the whole of the surface of the machine structure is subjected to vibratory motion (sound

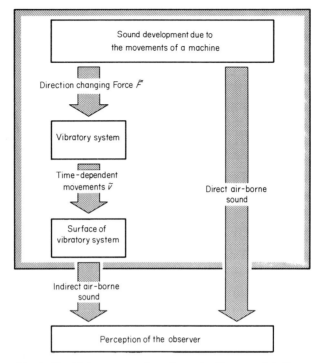

Fig. 2.63 Noise generation through direct and indirect air waves

conducted through solids) with a particular distribution of the vibrating velocity v_2. The vibration energy of the surface is converted into sound energy or sound pressure.

A sound-pressure pattern is thus established, according to the laws governing the propagation of sound and influenced by the space surrounding the

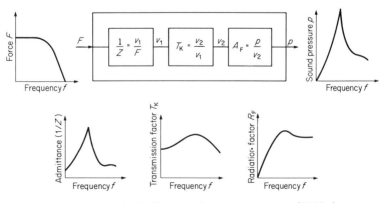

Fig. 2.64 Block diagram of noise generation (Lübke)

machine (see Volume 4). In addition to the individual contributory factors and the transmission blocks indicated in Fig. 2.64, the following generalized statements may be made.

Excitation force F(t). As the force amplitudes increase, so the sound pressure rises. The greater the force variation with respect to time (dF/dt) (i.e. the more pulsating the force pattern happens to be), the wider will be the range of the force excitation spectrum and hence the danger of large dynamic-vibration movements of the mechanical machine-component units.

Dynamic component characteristics. The input impedance

$$Z_E = \frac{F}{v_1} \qquad (2.37)$$

is the measure of the resistance of a mechanical structure to dynamic movements caused by the excitation forces. (The reciprocal value of the impedance is known as mobility.)

The transmission factor:

$$T_K = \frac{v_2}{v_1} \qquad (2.38)$$

is a measure of the transmission of the sound conducted through a solid between the point of excitation where the velocity is v_1 and the surface of radiation where the velocity is v_2.

From the input impedance Z_E and the transmission factor T_K we obtain the transmission impedance Z_T which gives the relationship between the excitation force F to the velocity v_2 for given positions of the radiating surface:

$$Z_T = \frac{F}{v_2} = Z_E \frac{1}{T_K} \qquad (2.39)$$

Dynamically stiff structures, i.e. those with a low resonance magnification (see sections 2.4 and 2.6.4), will inherently minimize the transmission of sound through the solid body, and consequently the vibration velocity and thus also the sound emission.

In order to avoid the transfer of the solid-body sound transmission from one machine unit to another, i.e. to minimize the transmission factor $T_K = v_2/v_1$, solid sound bridges must be removed. To this end, interfacings made from a springy soft material with high material-damping properties are used (e.g. rubber). In addition, changes in the shape of the cross-section and the introduction of a barrier mass will also reduce the velocity of the body sound waves (see Fig. 2.65). 'Solid sound barriers' is the term given to steps taken to prevent the spread of body sound by reflecting the body sound waves from elastic interfacings, changes in cross-section and barrier masses. The term 'solid sound damping' is used for the conversion of sound energy into

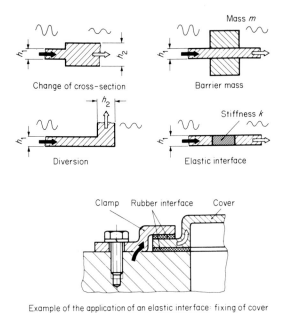

Fig. 2.65 Measures for reducing the transfer of solid-body sound transmission (solid sound barriers)

heat within the machine structure, e.g. material damping and damping at the joining faces (see sections 2.4 and 2.6.4).

Radiation. The conversion of mechanical vibration energy of the machine components into sound energy is determined from:

The radiation factor $R_F = \dfrac{p}{v_2}$ (Fig. 2.62) (2.40)

and

The radiation rate $\sigma = \dfrac{P_{sound}}{P_{vibration}} = \dfrac{P_{sound}}{Sv_2^2 \rho_L c_L}$ (2.41)

The logarithmic value of the radiation rate is indicated by $10 \log \sigma$.

Above a given limiting frequency f_0, which is determined from the longest flexural waves, the whole of the vibration energy is converted into acoustic energy ($\sigma = 1$; $10 \log \sigma = 0$).

The limiting frequency is given by

$$f_0 = \dfrac{c}{\lambda}$$ (2.42)

where c = velocity of sound in air
  λ = longest mechanical flexural wavelength of the mechanical structure

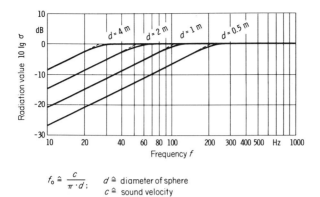

$f_0 \cong \dfrac{c}{\pi \cdot d}$; $d \cong$ diameter of sphere
$c \cong$ sound velocity

Fig. 2.66 Radiation values[22] as a function of component size, in this case of the spherical source of zeroth order and diameter

and when the radiation is taken from a sphere as the idealized body (Fig. 2.66):[22]

$$f_0 = \frac{c}{\pi d} \qquad (2.43)$$

Below this limiting frequency the radiation value rapidly declines, as shown in Fig. 2.66. Equation (2.43) and Fig. 2.66 indicate that the limiting frequency increases as the physical size of the machine component decreases, i.e. small machines radiate sound to a lesser degree in a lower frequency range than larger units.

Air-borne sound absorption and damping. Between the source of an acoustic air wave and the human ear, energy losses in varying degrees occur due to barrier and damping processes. By planned utilization of these effects, important noise-reducing measures are available.

Air-borne sound damping means the losses due to air friction and air sound absorption in walls. The sound energy which is absorbed in this manner is converted into heat. The absorption ratio is a measure of this phenomenon and is the relationship between the absorbed and the striking acoustic energy. Acoustically hard materials, i.e. smooth and hard walls, have a low absorption ratio. Walls covered with mineral wool or porous foam materials exhibit a relatively high absorption ratio (0.3 to 1.0 according to the frequency band).

Air-borne sound absorption is the resistance by a wall to the transmission of the sound. It is expressed in terms of the sound barrier ratio R. This is the logarithmic ratio of the striking acoustic power P_1 to the radiated acoustic power P_2 from the other side of the wall:[20]

$$R = 10 \log \frac{P_1}{P_2}$$

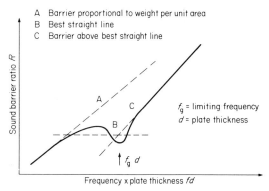

Fig. 2.67 Sound-barrier ratio in relation to frequency and plate thickness[23]

In general, the sound barrier ratio increases with rising frequencies and greater mass of the wall (thickness of plate d), as shown in Fig. 2.67.[23] However, when the sound is in the proximity of the plate's natural frequency in bending (range B in Fig. 2.67) the sound barrier may break down.

Both acoustic barriers and dampers are used to advantage when noise sources are encased[24] (see section 2.7.2) and when low-noise rooms are constructed by the utilization of acoustically damped walls and ceilings, as well as intermediate sound barrier boards.

2.7.2 Examples of noise reduction

One of the problems which hinder low-noise designs arises from the fact that detailed mathematical prognosis of the noise conditions of machines is not possible. This is due to the inadequate methods available for calculating the dynamic behaviour of a complex machine structure (see section 2.6.4) on one hand and to the multitude of unknown excitation sources on the other.

By knowledge of basic physical relationships (see section 2.7.1) and wide-reaching experience recorded in the form of case studies,[25,26] the designer nevertheless has the opportunity to take corrective action during the machine design stage, as well as subsequently when the machine is in service. An effective reduction in the noise output of a machine is usually the result of many individual and painstaking improvement efforts. Consequently, it is important to know what effect any single removal of a noise source will have on the total noise emission of a machine.

Figure 2.68 illustrates this relationship.[18] Suppose, for example, that there is a transmission level difference ΔL of 4 dB between the residual sound level of the machine L_1, which is radiated without emission from the source to be quieted, and the acoustic level L_2, which is given out by that source itself. A successful noise reduction at the acoustic source L_2 of 6 dB will only result in a 1.1 dB reduction in the total machine acoustic level. The same diagram may

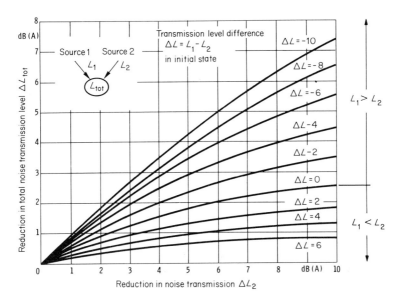

Fig. 2.68 Effect of reduction of noise transmission from one source on the total noise transmission

be used when planning the layout of a machine shop, i.e. to determine the contribution to the total noise level by a particular machine.

Knowledge of this mathematical relationship enables an estimate to be made of the success which may be expected when particular noise-reduction steps are about to be taken and prevents disappointments which can frequently be forecast when costly noise reduction measures lead to hardly measurable changes in the total machine noise level. In accordance with the relationships described in section 2.7.1, noise-reduction measures may be applied at any point of the noise chain, from the point-of-force application right up to the acoustic pressure on the ear. A decision must be made in a given case to determine at which point in that chain useful action may be taken.

Figure 2.69 categorizes noise reductions into active and passive types, which are then further subdivided into primary and secondary measures. Active noise reductions are directed towards the applied force and the bodily sound generation (primary), as well as to that of transmission and radiation (secondary). Passive measures are intended to influence the intensity of the sound, i.e. by means of special sound barriers such as encasements (primary) and by paying attention to the acoustics of the machine shop and/or the provision of ear defenders (secondary).

Examples are given for the first three types of measures mentioned in the lower block diagram of Fig. 2.69. (Further examples can be found in references 20, 21 and 24 to 26).

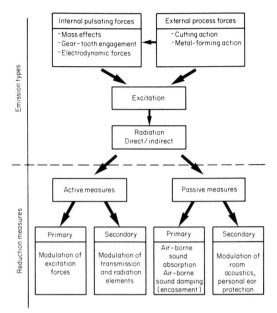

Fig. 2.69   Noise emission and measures for their reduction

2.7.2.1   *Active–primary measures*

These are particularly necessary in geared drive units. The vibration excitation in gear drives occurs in a variety of manners. Gear-tooth form of errors cause uneven torque transmission and pulsating tooth loading. Consequently, high-quality gear cutting is a basic requirement for quiet-running drives, but even faultlessly produced gear wheels tend to produce dynamic excitations under load upon the whole structure (wheels, shafts, casings).

As may be noted from Fig. 2.70, the stiffness of the mating gear teeth is dependent upon the rolling angle. In addition, according to the degree of overlap, continuous reversals between single and double contact introduce large variations in stiffness which can initiate considerable vibratory effects. To minimize this phenomenon, the tooth geometry should be such that the degree of overlap is over two or three teeth (by using long addendum or helical teeth). In this way, the stiffness variations are much reduced.

Another major force excitation results from the elastic deformation of the gear teeth under load. This has the effect of making the angular motion of the driving gear-wheel body slightly greater than the theoretical angle, the driven wheel having a smaller movement by the same amount. The teeth which are about to come into contact and are still load-free will therefore experience interference to a degree dependent upon the load torque. As this prevents proper meshing, the teeth are also distorted as they take up the load (Fig. 2.70).

71

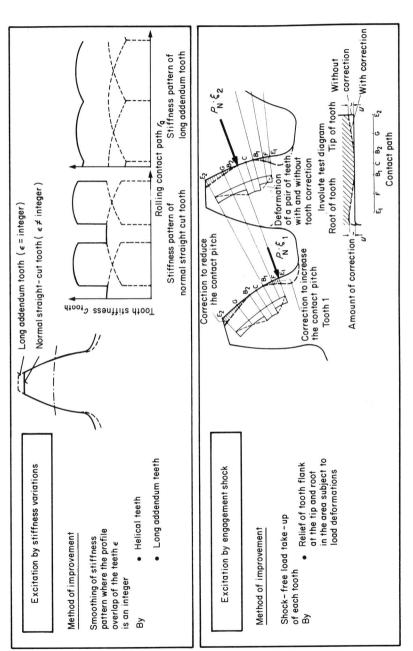

Fig. 2.70 Active-primary measures to reduce noise emission in gear drives

In order to minimize the sudden shock load and to enable the incoming pair of teeth to take up the load smoothly and steadily, the involute of the addendum of the driven teeth is modified by the degree of tooth deformation, as shown in Fig. 2.70. However, it must be borne in mind that these measures will reduce the degree of profile overlap. In order therefore to minimize the stiffness reduction, complicated and costly design calculations are necessary.

2.7.2.2 Active–secondary measures

The damping of a mechanical structure from which noise radiates is not always a simple matter. As an example, a number of different measures are shown in Fig. 2.71 as applied to circular-saw cutters which are particularly prone to vibrations resulting from the applied cutting force. The success of these damping efforts is considerable.[27]

2.7.2.3 Passive–primary measures

In many instances, partial or even total encasement is the only practical and economic answer. Figure 2.72 illustrates the shrouding of a motor. The wall of the encasement is usually constructed in such a way that a damping layer of mineral wool absorbs a part of the sound energy. In order to minimize the

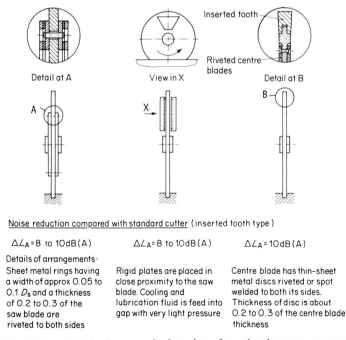

Noise reduction compared with standard cutter (inserted tooth type)

$\Delta L_A = 8$ to $10\,dB(A)$ $\quad\quad$ $\Delta L_A = 8$ to $10\,dB(A)$ $\quad\quad$ $\Delta L_A = 10\,dB(A)$

Details of arrangements:

Sheet metal rings having a width of approx 0.05 to 0.1 D_s and a thickness of 0.2 to 0.3 of the saw blade are riveted to both sides

Rigid plates are placed in close proximity to the saw blade. Cooling and lubrication fluid is feed into gap with very light pressure

Centre blade has thin-sheet metal discs riveted or spot welded to both its sides. Thickness of disc is about 0.2 to 0.3 of the centre blade thickness

Fig. 2.71 Solid-body sound damping for circular-saw cutters (Westphall)

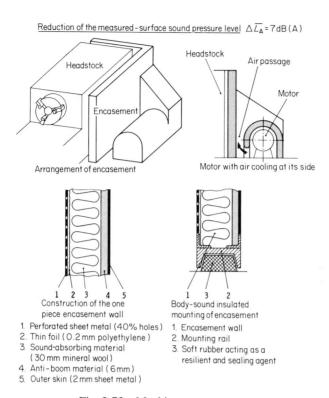

Fig. 2.72 Machine encasement

penetration, i.e. to provide the greatest possible sound barrier, the outer skin should be as massive (thick walled) as possible. This, however, results in the need for heavy and expensive casings.

A compromise is the solution shown in Fig. 2.72. The thin (2 mm) steel sheet metal is covered with a 6 mm thick sound-absorbing layer of floating plastic material having high inherent damping properties. This covering provides the necessary mass whilst at the same time drastically reducing the vibration amplitudes of the natural frequency of the casing due to the high damping property of the material. The isolation of the casing from body vibration is of great importance, so that only the air sound waves are received and absorbed.

Any apertures in the casing required for functional purposes, and which would allow the sound to escape, must be kept as small as possible. The effectiveness of the encasement is greatly reduced by such openings and therefore they should be avoided wherever possible.

3

INSTALLATION AND FOUNDATIONS OF MACHINE TOOLS

The dependence of a machine tool's static and dynamic behaviour on the foundations and the mounting elements upon which it stands varies with the type of machine. There are three main factors which must be considered when designing machine foundations and installations:

(a) levelling and alignment of the machine;
(b) additional stiffness for the machine supplied by the foundations;
(c) active and passive isolation against dynamic disturbances.

Whether there is a need for the machine mountings to provide insulation or additional stiffness, or both of these features, is above all dependent upon the type of machine, i.e. its stiffness characteristics, its applications and its expected degree of accuracy, as well as upon its location. Consequently, considerable expenditure may be necessary for the provision of adequate foundations and mounting elements.

The methods of installing machine tools may be classified into four main groups, as shown in Fig. 3.1.

The first category consists of smaller machine tools such as lathes, milling machines and shaping machines, which have adequate inherent rigidity and do not require to be further stiffened by means of their foundations.

When positioning such machines on the workshop floor or on shock-absorbing surfaces, it is advisable to guard against foreign or reactive dynamic interference by an elastic mounting, combined with active or passive isolation. As there is a wide range of low-cost isolation mounts on the market—steel and rubber sprung—such a mounting procedure is always advantageous (see column 1 in Fig. 3.1).

Similar installation procedures, but with considerably greater costs due to the active isolation, are utilized for metal-forming machines, such as forging hammers, presses and shearing machines. In these instances, it is necessary to absorb the impact energy and to minimize the forces transmitted into the floor by a soft springing system, which will provide the lowest possible natural

Machine tools classified according to their installation					
Type of machine	Small machines: Lathes Milling machines Shaping machines	Forming machines: Presses Blanking presses Forging hammers	Precision machines: Roller grinding machines Precision turning lathes	Medium and heavy machine tools: Planer milling machines Drilling and milling centres	
Rigidity of machine	Adequate stiffness	Good stiffness	Inadequate stiffness	Inadequate stiffness	
Loading of mounting elements	Minimal loading	Very heavy impact loading	High static loading	Medium to high static loading	
Functions of mounting elements	Active and passive isolation alignment	Active isolation	Passive isolation	Stiffening alignment	
Mounting criteria	Dynamic	Dynamic	Dynamic	Static	Static
Machine				X	X
Mounting elements	X	X	X	X	X
Foundation		X	X	X	X
Soil	X	X			X

Fig. 3.1 Criteria for the installation of machine tools

frequency for the 'machine-mounting element' system. In particularly critical instances, double-foundation elements are utilized, consisting of a foundation block as a seismic smoothing mass and a foundation trough as support against the floor and spring and damping elements. In relation to the machine costs, the necessary expenditure for such mounting methods is very high (see column 2 in Fig. 3.1).

In the case of precision machines such as roller grinding machines and precision turning lathes, a similarly costly foundation system is necessary in many instances. As such machines are required to produce highly accurate products and very good surface finishes, but usually do not have the necessary inherent rigidity, a sound foundation must provide passive isolation, as well as additional rigidity. The foundation block, which is supported by a system of springs against the floor, increases the total mass; this has a favourable effect as it helps to reduce the frequency of the system and in addition assumes the role of a stiffener through the frictional connection with the machine (see column 3 in Fig. 3.1).

The installation of medium-sized and most heavy machine tools from the rigidity viewpoint constitutes a further category. Mainly due to economic considerations, machine beds and mounting units of planer milling machines and drilling and milling centres cannot be provided with adequate built-in stiffness. The foundation unit which is set into the soil of the shop floor must provide a predetermined bending and torsional rigidity, so that the machine can perform to the required specification, notwithstanding the effects of its own weight and the high loads induced by the production process. The floor itself must have an inherently adequate load-carrying factor and solidity to avoid the effects of settlement on the foundations and the machine frame. Active or passive isolation of such machines is desirable in many instances, but is not practical owing to the very high costs involved. A compromise is obtained by the separation of the machine foundations from each other and their neutralization from the shop floor (see column 4 in Fig. 3.1).

3.1 Installation and foundations of machines with adequate inherent rigidity

In the case of small, inherently rigid machines, such as, for example, lathes, milling machines and shaping machines, the foundations are not situated directly in the force flux flow of the static applied working forces. However, the mounting conditions have in such instances a greater or lesser influence upon the dynamic behaviour at the cutting point (workpiece–tool).

As an example of such conditions, Fig. 3.2 shows the amplitude–frequency characteristics of the relative receptance (i.e. the dynamic equivalent of the static flexibility) at the cutting point of a chucking lathe, due to the excitations and distance travelled in the x axis, for differing mounting conditions.

The fact that the machine has inherent stiffness is responsible for the static resilience (when $f = 0$ Hz) being virtually independent of the mounting conditions at $d = 0.04$ μm N^{-1}. It may also be noted from the curves that the dynamic behaviour is influenced by the mounting methods throughout almost the whole frequency range. The maximum value of the dynamic receptance reduces from 0.15 μm N^{-1} when rigidly mounted to 0.1 μm N^{-1} when soft foundation elements are employed. The type of installation (i.e. soft or damped/hard) enables variations in the dynamic behaviour of inherently rigid machines to be achieved.[28]

The relationship of the dynamic behaviour at the cutting point of the machine to the method of mounting is influenced by the mode shape of the vibration and the consequential enforced movements at the mounting support

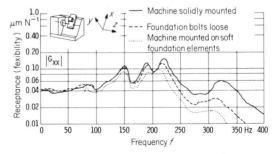

Fig. 3.2 Influence of differing mounting conditions on the dynamic behaviour of a chucking lathe

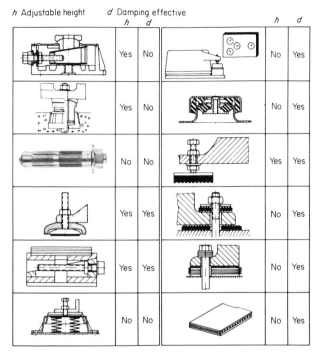

Fig. 3.3 Varying machine mounting elements

points. A meaningful distinction between the advantages and disadvantages of varying methods of mounting (soft or solid) can only be made in relationship to the characteristics of the machine involved. A number of mounting elements are illustrated in Fig. 3.3. According to their method of construction, height adjustments of the elements may be possible. They range from simple damping plates to adjustable units with spring and damping characteristics.

A simplified model in the form of a single-mass vibrator may be used for the construction of the vibration-isolated installation of inherently stiff machines. In this approach, m represents the mass of the machine, k the resilience and c the damping coefficient of the mounting elements. Depending upon the particular case, an installation may be constructed with the aid of the relationships depicted in Fig. 3.4[29].

In the case of active isolation the elastic foundation is required to minimize floor vibrations caused by the dynamic forces of the machine (F_0). Figure 3.4 indicates the transmissibility, i.e. the ratio of the force activated upon the floor (F_A) to the excitation force (F_0) caused by the working process plotted against the relative excitation frequency (ω/ω_0) and the damping factor (D). Passive isolation with the aid of damped elastic machine mountings is used when it is desired to minimize the effect of any floor vibrations which are present upon the production machine. To this end, Fig. 3.4 also shows the

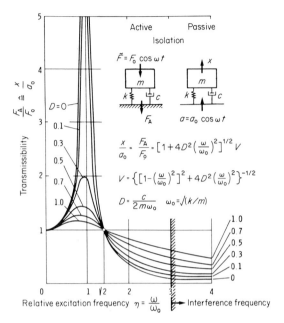

Fig. 3.4 Effects of active and passive vibration isolation

relationship of the machine-motion amplitudes (x) to the floor-motion amplitudes (a_0).

As may be seen from the diagram, the natural frequency of mountings is given by

$$\omega_0 = \sqrt{\left(\frac{\Sigma k}{m}\right)} \qquad (3.1)$$

which should be kept as low as the spring stiffness (k) of the mounting elements allows. It should be less than one-third of the interference frequency (Fig. 3.4). Lightly damped mounting elements exhibit better isolation and barrier effects in the overcritical frequency range, but may cause interfering, lifting or pitching vibration movements of the whole machine.

As a guide for the damping factor the following may be stated:

$$0.4 < D = \frac{\Sigma c}{2m\omega_0} = \frac{\Sigma c}{2\sqrt{(m\,\Sigma k)}} < 0.7 \qquad (3.2)$$

3.2 Installation and foundations of metal-forming machines

In the case of metal-forming machines it is of the utmost importance to minimize disturbing effects upon the environment caused by the pulsating

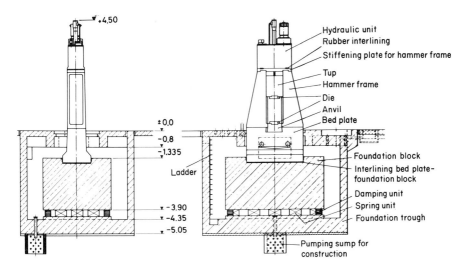

Fig. 3.5 Main constructional elements of sprung mountings for a forging hammer

working forces. The energy transmitted by the machine into the surrounding area (floor hammering) must be kept to a minimum.

An example of an actively damped foundation mount for a forging hammer is shown in Fig. 3.5. The system is activated when the tup drops on to the bed plate, on top of which the dies and work are placed. The bed plate of the forging hammer lies on a rubber interlining and the foundation block, which in turn is supported from the floor of the foundation trough by springs and damping units. With an optimum layout of such an arrrangement, as in Fig. 3.4, the movement of the floor can be less than 10% of the bed plate motion. The stiffness of the spring elements basically governs the movement of the floor (vibration), whilst the damping characteristics control the decay conditions. The mass of the foundation performs the function of increasing the total mass and thus the barrier effect.

The machine mounting units which are available today enable drop forging hammers to be placed directly on the shop floor (i.e. without the need for an additional mounting mass), as shown in Fig. 3.6. Whilst this dispenses with the need to provide expensive vibration foundations, it has the disadvantage of possibly inducing larger vibration movements of the machine.

Since the passing of special legislation in the Federal German Republic,[30] particular attention must be paid to the soil vibrations of metal-forming machines. As a basis for quantifying the tremors, the 'curves of equal perception strength' are applied, which are presented and defined in the draft standard DIN 4150[31] and the draft VDI 2057.[22] In VDI 2057, differentiation is made between various body attitudes (undefined, standing, lying and seated) and force applications (horizontal and vertical). In general terms, the tremor factor is defined as the K value.

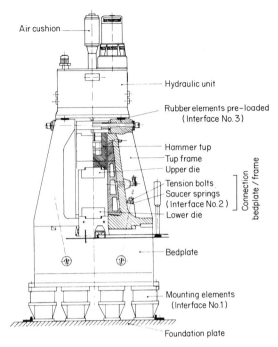

Fig. 3.6 Principal construction of a directly sprung forging hammer (Lasco)

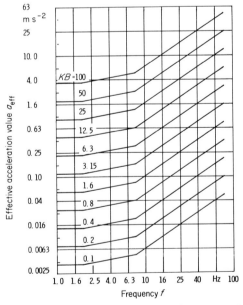

Fig. 3.7 Curves of equal perception strength KB for undefined body attitude (VDI 2057)

Figure 3.7 depicts as an example the curves of equal perception strength *KB* for an undefined body attitude as a function of the effective value of the acceleration and excitation frequency.[22] These correspond to the evaluation technique given in DIN 4150.

In practice, a mixture of vibrations is usually experienced. With the application of a Fourier analysis, the acceleration amplitude of each frequency may be determined. The effective total tremor factor is obtained from the *K* values of the individual frequencies dependent on the acceleration amplitudes:

$$K_{tot} = \sqrt{(K_1^2 + K_2^2 + \cdots + K_n^2)}$$

If the tremor factor and the time of activity are known, then an evaluation may be made with the use of the curves shown in Fig. 3.8. Limiting values are defined for the tremors directly attributed to the metal-forming machine for the impairment of well-being, performance and health.[22]

The tremors travel through the soil and in some instances may excite strong vibrations in buildings at some considerable distance from the source, e.g. in floors and ceilings. In the specification DIN 4150, the criterion used for evaluation is the 'disturbance to the participating persons'. Figure 3.9 gives the reference values for evaluation. The classification of the building areas is governed by utilizaton of the buildings. Day is defined in the standard as between 6.00 and 22.00 hours, unless special local regulations apply. The values applicable to metal-forming machines are those in column 3 of Fig. 3.9. The figures stated in brackets are to be used for vibrations of <5 Hz.

As a first approximation for an actively isolated vibration installation, the metal-forming machine with its mounting element may be considered as a

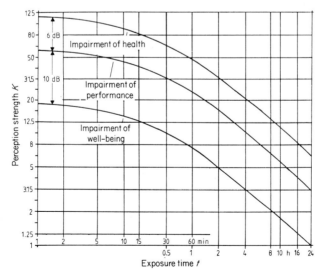

Fig. 3.8 Limiting curves of perception strength *K* in relation to exposure time *t* (VDI 2057)

	1	2	3	4
			KB- limiting values	
	Type of area	Time	Tremors which are continuous and repeated with pauses	Rarely occurring tremors
1	Purely residential area (WR) Generally residential area (WA)	Day	0.2 (0.15)	4
	Weekend housing area (SW) Small housing estate (WS)	Night	0.15 (0.1)	0.15
2	Village area (MD) Mixed area (MI)	Day	0.3 (0.2)	8
	Inner city area (MK)	Night	0.2	0.2
3	Business area (GE) (Also valid for offices)	Day	0.4	12
		Night	0.3	0.3
4	Industrial area (GI)	Day	0.6	12
		Night	0.4	0.4
5	Special types according (SO) to utilization and proportion of residences	Day	0.1 to 0.6	4 to 12
		Night	0.1 to 0.4	0.15 to 0.4

Fig. 3.9 Reference values for the evaluation of tremors in living accommodation and comparable rooms (DN 4150)

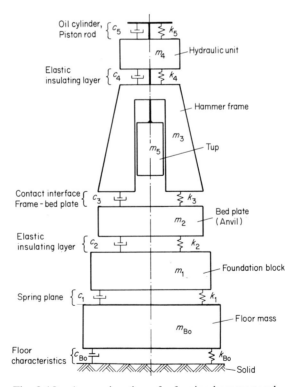

Fig. 3.10 Approximation of a forging hammer and its foundations and floor as a six-mass vibrator

single-mass vibrator (see Fig. 3.4). Even in such a case, the aim should be to obtain a resonance frequency (ω_0) of the complete system which is low in relation to the exciter frequency (ω). In the case of presses it is particularly important to note that the complete system will additionally be subjected to tilting oscillations.

For a more precise calculation of the spring in the foundations of the complete system, the model of a multi-mass vibrator shown in Fig. 3.10 may

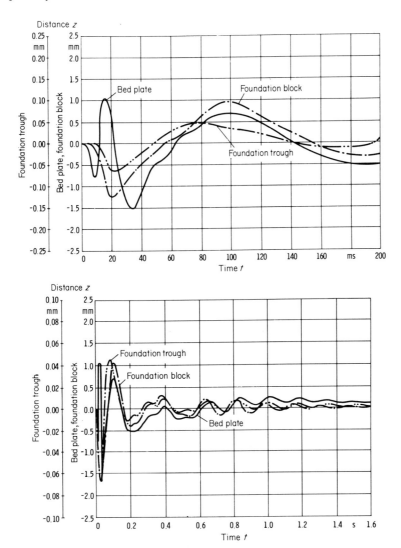

Fig. 3.11   Time-based progression of fade-out after impact of bed plate, foundation block and foundation trough (actual)
   Time span 200 ms   (upper diagram)
   Time span 1.6 s   (lower diagram)

be considered. In this case the individual components of the structure are considered to be rigid, and are connected to each other by 'no-mass' spring and damping elements.

Figure 3.11 shows the displacements which occur on each component of the forging hammer system after one stroke of the tup. In the upper graph a time span of 200 ms is considered, and it is clear from the lower graph that the system is stationary after about 1.6 seconds. The installation of presses on which the load peaks are smaller than on forging hammers is also undertaken largely from the viewpoint of minimal environmental disturbance. Here, too, an active isolation from the neighbouring floor area is provided.

3.3 Foundations for precision machines without adequate inherent rigidity

In the case of precision production machines with inadequate inherent rigidity, such as roller grinding machines, precision turning lathes, etc., the foundations perform the dual function of providing additional stiffness for the machine, on the one hand, and isolation from external dynamic interferences, on the other.

Figure 3.12 gives an example of a foundation for a roller grinding machine. In order to minimize the effect of external interferences, a passive isolation is provided, i.e. the foundation is laid directly into the floor but is supported on steel springs and viscose dampers. As support against the ground, the sprung foundation block is placed into a steel reinforced-concrete trough.

The sprung foundation block provides additional rigidity for the machine frame. The distortions due to the weight of the work and the operation of the slide are required to be as small as possible. Furthermore, the mass of the foundation must be sufficiently large so that, when added to the mass of

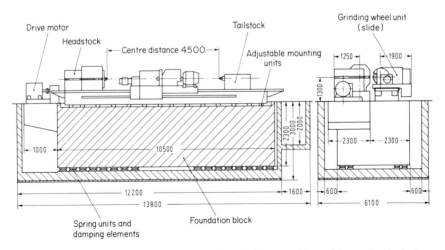

Fig. 3.12 General arrangement of a roller grinding machine with passive isolation (Waldrich-Siegen)

the machine, an adequate absorber mass against external disturbances is available.

With regard to external excitations, this system can also be considered for a first approximation as a single-mass vibrator. The transmission characteristics of the machine's movements are illustrated in Fig. 3.4 in relation to the floor vibration amplitude and the stiffness and damping characteristics of the mounting elements.

3.4 Foundations for medium and heavy machine tools without adequate inherent rigidity

In the case of heavy machine tools which have inadequate inherent rigidity, the foundations exclusively serve to improve stiffness. When such machines are installed on their foundations, rigid mounting elements (wedges) are basically used.

Figure 3.13 shows a diagram of a gantry milling machine with its foundation. The size and shape of the upper surface of the foundation are determined by the external dimensions of the machine, as well as by some design considerations such as swarf disposal, cable channels, etc. The thickness of the foundation block, however, is dependent upon the soil conditions and the required overall rigidity, which the expected performance of the machine demands. When the foundation block is designed, it is desirable that the calculations are based upon the whole system, consisting of the machine, installation units, the foundation block interface elements and the floor itself.

A simplified mathematical model is shown in Fig. 3.14, which does *not* take account of the stiffness characteristics of the machine itself or of the interfacing elements between the machine and the foundation block. The weights of

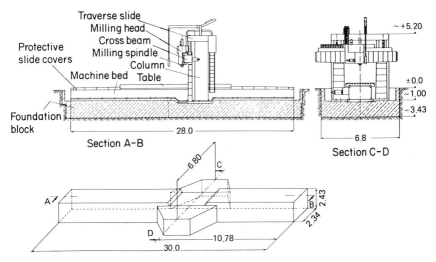

Fig. 3.13 Gantry milling machine, flat based

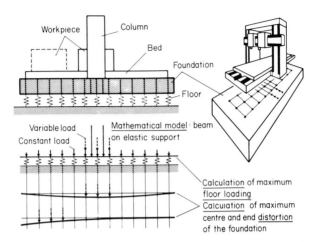

Fig. 3.14 Static calculations for machine-tool foundations. Beam on an elasticated (sprung) support

the machine and of the work are considered in the form of loads acting directly on the foundation block. The foundation block itself is approximated as a beam and the floor is simulated by springs. The load per unit area on the floor is calculated and the resultant bending of the beam for a given load is obtained (see the lower part of Fig. 3.14).

A more precise design of the foundation is possible by applying the mathematical model shown in Fig. 3.15.[32] In this method, the machine characteristics and the mounting elements are included in the calculations. The machine is approximated by a beam and the interfacing elements by springs; the foundations are also approximated by a beam and the floor by compression and rotary springs. As the cutting forces are only of secondary import-

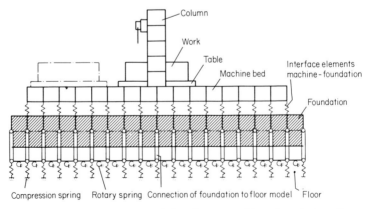

Fig. 3.15 Mathematical model of a machine–foundation–floor system for determination of the distortion characteristics

ance as far as the foundations are concerned, only the weights of the machine itself and that of the work are taken into account, even in this method.

The calculations are based upon the finite element mathematical method (see section 2.6.2).

When the simplified method illustrated in Fig. 3.14 is used for the design, a larger foundation is usually built than the minimum size necessary, because the rigidity of the machine has not been taken into account. Consequently, the costs are greater than need be. Hence the more exact solution defined in Fig. 3.15 will prove to be more economic in spite of its higher computation costs,[33] as considerable material and construction costs are likely to be saved when the foundations are built.

4

GUIDEWAYS AND BEARINGS

The guideways or guides which are used for the movement of slides and work tables and the bearings which support the main spindles are amongst the most important construction units which lie in the force-flux flow of a machine tool.

Guideways and spindle bearings are required to:

(a) accurately control the motion of the tool and work, while at the same time
(b) absorb all external forces (process forces and weight loading).

Deviations from the required feed or cutting motions reduce the accuracy of the finished product.

Out of the three linear and three rotary degrees of freedom, at least four—in most cases, five—are restrained in a moving element, depending upon the function it is required to perform (Fig. 4.1). For example, a slide moving in a straight line is left with only one linear degree of freedom.

A drilling spindle in the spindle guide of a drilling machine retains one linear and one rotary degree of freedom, whilst the main spindle of a lathe has only one rotary degree of freedom remaining.

Arising from the degrees of freedom which remain on a guided machine-tool element, one may derive a first classification of guides based upon the *type of movement*: linear, rotary or a combination of these two. A further differentiation may be made by considering whether the slides or bearings are moved during the actual production process. If such movement does occur, then they are referred to as *working guideways*, and if there is no movement during machining, then the term *setting guideways* may be used. Such setting guides are used for the positioning of a machine element before and/or after the actual working process. Generally, such elements are securely clamped on their guideways during machining. However, in a number of instances a guideway is required to perform functions according to the particular requirement at hand, i.e. it must operate as a working guide in one instance and as a setting guide in another (see Fig. 4.2).

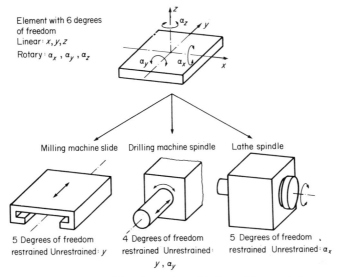

Fig. 4.1 Degrees of freedom of a guided machine tool element

The contact zones of working guideways are required to have greater performance capabilities in regard to their lubrication and wear resistance because the movements are under load.

The working guideways may be further subdivided into plain or rolling guides and bearings. The elements which move relative to each other in plain

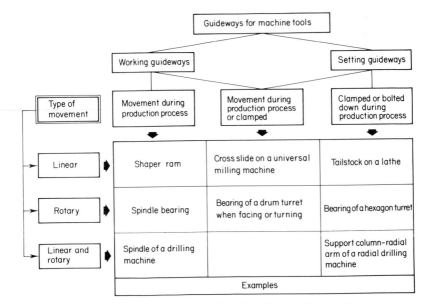

Fig. 4.2 Guideways and bearings for machine tools

guides or bearings are partially or wholly separated from each other by a lubricating film. In the case of rolling bearings or guides, such separation is provided by the rolling element. The type of lubrication medium and method of build-up of the lubricating film allows a differentiation between the following types of guides and bearings (Fig. 4.3):

(a) hydrodynamic,
(b) hydrostatic,
(c) aerodynamic and
(d) aerostatic.

In the case of hydrodynamic plain slides and bearings the lubricating media is applied without, or with minimal, pressure into the contact area between the slide and bed or the shaft and bearing. The lubricating film is built up automatically due to the relative movement of the slide on the slideway. This also applies to aerodynamic plain slides and bearings, where air serves as the 'lubricating medium'. On the other hand, hydrostatic and aerostatic units have the lubricating film (oil or air) supplied and maintained by an external pressurized system.

Irrespective of the guiding principle being used, guideways and main spindle bearings of machine tools have varying needs for:

(a) rigidity;
(b) damping capabilities normal and parallel to the guided surface;
(c) geometric and kinematic accuracy;
(d) low frictional forces;
(e) wear resistance;
(f) low play or freedom from play;
(g) provision for adjustment of play.

Further, when guideways are designed and constructed, consideration must be given to:

(a) favourable positioning in relation to the work area;
(b) protection against swarf and damage;
(c) unrestricted swarf disposal;
(d) positioning of the feed force near the line of the resultant force arising from the process, frictional and mass forces.

These criteria vary in importance depending upon a particular application, and hence the selection of the design for the guides and their geometry is in some cases quite critical. Figure 4.4 compares the main characteristics of various types of main spindle bearings. These differing features are discussed in detail in the following sections.

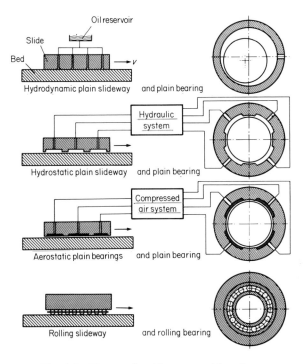

Fig. 4.3 Types of guideways and bearings

Characteristic	Hydrodynamic bearing	Rolling bearing	Hydrostatic bearing
Damping	●	○	●
Running accuracy	●	◐	●
Speed range	○	◐	●
Wear resistance	◐	◐	●
Power loss	●	○	◐
Installation costs	◐	○	●
Cooling capacity	◐	◐	●
Reliability	●	●	◐

Evaluation of characteristics:
● high　　◐ medium　　○ low

Fig. 4.4 Characteristics of various types of bearings for main spindles

4.1 Hydrodynamic plain guideways and bearings

The most widely applied type of guiding system in machine tools is the hydrodynamic plain guideway. The most important reason for this is the low manufacturing cost and at the same time the good performance when compared with other designs. Modern methods of production in conjunction with the application of plastic materials contribute to this advantage. On the other hand, hydrodynamic lubricated main spindles are much rarer today; their main application is to be found in precision manufacturing machines, such as lathes and grinding machines, and in the heavy machine-tool construction field (presses). From certain points of view, the performance of hydrodynamic plain bearings is superior to other types of bearings, and this will be discussed in detail later. Firstly, a number of basic principles of hydrodynamic lubrication theory and frictional behaviour are discussed, which are of importance in the action of plain guideways and bearings.

4.1.1 Fundamentals of friction and lubrication

4.1.1.1 The concept of viscosity

Viscosity is the physical property of the internal molecular frictional resistance of a fluid.[34] If two surfaces have a fluid film between them, as shown in Fig. 4.5, and a relative motion occurs between such surfaces, then the cross-section of the fluid is subjected to shear. According to Newton's law, the shear stress in the cross-section of a fluid is proportional to the velocity gradient dv_s/dy of the flow in that cross-section. The constant of proportionality between the shear stress τ and the velocity gradient dv_s/dy is the viscosity η, also known as dynamic viscosity:

$$\tau = \eta \frac{dv_s}{dy} \tag{4.1}$$

In a linear velocity profile, as in Fig. 4.5, this may be expressed as

$$\tau = \eta \frac{v}{h} \tag{4.2}$$

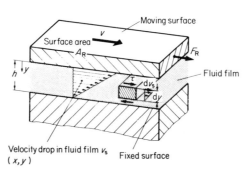

Fig. 4.5 Schematic representation of the shearing of a fluid film

The force required to move a body having a frictional surface area A_R with velocity v over a fluid film is given by

$$F_R = \tau A_R \qquad (4.3)$$

Substituting from equation (4.2) into (4.3) we get the general expression for the frictional force:

$$F_R = \eta \frac{v}{h} A_R \qquad (4.4)$$

Correspondingly, the frictional energy which may be converted into heat in the fluid film will be

$$P_R = F_R v \qquad (4.5)$$

or

$$P_R = \eta \frac{v^2}{h} A_R \qquad (4.6)$$

These equations only apply to the so-called Newtonian fluids. They are not valid, for example, for greases and pastes. Figure 4.6 indicates the relationship between shear stress and shear velocity of Newtonian and non-Newtonian fluids.[34]

The strong dependence of the viscosity of an oil upon temperature is of particular importance when the performance of hydrodynamic guides is considered. Hence, the viscosity value of a lubricating oil is only meaningful when related to a temperature. Figure 4.7 illustrates the relationship between viscosity and temperature for four different lubricating oils.

The unit in which dynamic viscosity is expressed is

$$\eta \quad (\text{N s cm}^{-2}) \quad \text{at } T \quad (K) \qquad (4.7)$$

Very often, the old unit poise (P) and centripoise (cP) are still in use today. The conversion from Poise is obtained from:

$$1P = 100 \text{ cP} = 10^{-5} \text{ N s cm}^{-2} \qquad (4.8)$$

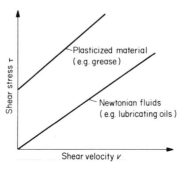

Fig. 4.6 Shear characteristics of Newtonian fluid and plasticized materials (Fuller[34])

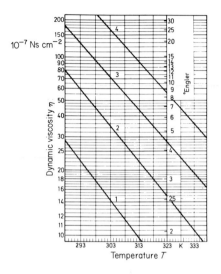

1. Low viscosity spindle oil for high speed spindles
2 and 3 Spindle and gear box oil
4 High viscosity oil for plain guides

Fig. 4.7   Dynamic viscosity η relationship to temperature

The unit 'degree Engler' (°E) is still occasionally in use, but there is no linear relationship between this unit and newton-seconds per square centimetre (see Fig. 4.7). This is because a different method is used to measure the viscosity in degrees Engler.

If the dynamic viscosity is considered in terms of the density of the lubricating fluid, then the dynamic viscosity μ is obtained, i.e.

$$\nu = \frac{\eta}{\rho} \qquad (4.9)$$

The unit for the kinematic viscosity μ is square centimetres per second (cm² s⁻¹). Here, too, the old units are often used for kinematic viscosity, viz. stoke (St) and centistoke (cSt). There is no numerical difference here as

$$1 \text{ St} = 1 \text{ cm}^2 \text{ s}^{-1} \qquad (4.10)$$

In order to appreciate the magnitude of viscosities of different fluids, a number of examples are quoted in Table 4.1.

4.1.1.2   Hydrodynamic pressure build-up

If, in contrast to the conditions depicted in Fig. 4.5, a flat body moves over a wedge-shaped lubricating fluid film, a fluid pressure occurs due to the pull of the fluid into the gradually reducing wedge space. This is capable of support-

Table 4.1 Viscosities of different fluids at 294 K (21 °C)[34]

Fluid	Dynamic viscosity η (N s cm^{-2})
Heavy motor oil SAE 50	800×10^{-7}
Glycerin	500×10^{-7}
Medium motor oil SAE 30	300×10^{-7}
Winter motor oil SAE 10	70×10^{-7}
Extra light motor oil SAE 5	32×10^{-7}
Ethylene glycol	20×10^{-7}
Mercury	1.5×10^{-7}
Turpentine	1.45×10^{-7}
Water	1.0×10^{-7}
Air	0.018×10^{-7}

ing a load and thus causes the body to float. Figure 4.8 illustrates diagrammatically the conditions which exist in such a fluid wedge.

A sliding flow is superimposed due to the drag of the oil particles by the moving surface, travelling at velocity v, and the pressure flow caused by the reducing gap in the wedge. This induces flow velocity patterns $v_s(x,y)$ as indicated on the diagram. Above the fluid wedge the pressure distribution exists as shown. In the following paragraphs a brief description is given of the physical conditions which exist in the wedge.

Let it be assumed that the wedge space is infinitely wide in the direction normal to the x, y plane, i.e. there is no flow of the fluid in that direction.

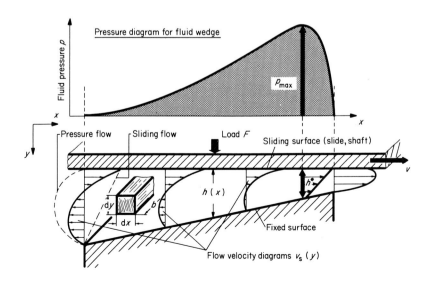

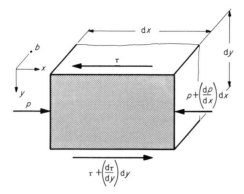

Fig. 4.9 Compression and shear forces acting upon a small section of fluid in a wedge space

Compression and shear forces are acting on the small section of the fluid $dx\,dy\,b$, as shown in Fig. 4.9.

When the summation equation of the forces acting in the horizontal direction is formulated, then for conditions of equilibrium we have:

$$\frac{dp}{dx} = \frac{d\tau}{dy} \quad (4.11)$$

When differentiating equation (4.1) we get:

$$\frac{d\tau}{dy} = \eta \frac{d^2 v_s}{dy^2} \quad (4.12)$$

From equations (4.11) and (4.12) we have:

$$\frac{d^2 v_s}{dy^2} = \frac{1}{\eta}\frac{dp}{dx} \quad (4.13)$$

By partially integrating equation (4.13) twice with respect to y, we obtain the velocity distribution in direction y (dp/dx is constant for y):

$$v_s(y) = \frac{1}{2\eta}\left(\frac{dp}{dx}\right)y^2 + C_1 y + C_2 \quad (4.14)$$

C_1 and C_2 are obtained from the two boundary conditions:

When $y = 0$ then $v_s(y) = v \rightarrow C_2 = v$
When $y = h(x)$ then $v_s(y) = 0 \rightarrow C_1 = -\dfrac{v}{h(x)} - \dfrac{1}{2\eta}\dfrac{dp}{dx}h(x)$ $\quad (4.15)$

From this, the flow velocity is obtained:

$$v_s(y) = \underbrace{\frac{1}{2\eta}\frac{dp}{dx}\left[y^2 - h(x)y\right]}_{\text{pressurized flow}} + \underbrace{v\left[1 - \frac{y}{h(x)}\right]}_{\text{sliding flow}} \quad (4.16)$$

Equation (4.16) presents the velocity distribution in the fluid film as a parabolic pressure flow with the linear sliding flow superimposed (see Fig. 4.8).

With regard to the pressure curve, and considering the point of maximum pressure, it may be shown that this maximum occurs at the smallest gap in the wedge. At this point the pressure differential $dp/dx = 0$. It follows from equation (4.16) that the flow-velocity pattern at the point of maximum pressure where $p = p_{max}$ becomes:

$$v_s(y) = v\left(1 - \frac{y}{h^*}\right) \quad (4.17)$$

where h^* is the wedge gap at the point of maximum pressure.

The flow quantity Q through any cross-section of the wedge space, using equation (4.16), is given by:

$$Q = \int_{y=0}^{y=h(x)} v_s(y) b \, dy \quad (4.18)$$

Also, the flow quantity at the point of maximum pressure can be stated from equation (4.17) as:

$$Q^* = \frac{v}{2} bh^* \quad (4.19)$$

From equations (4.16) and (4.17) and assuming a steady flow we get:[34]

$$\frac{dp}{dx} = 6\eta v \left[\frac{h(x) - h^*}{h(x)^3}\right] \quad (4.20)$$

With the use of equation (4.20), the pressure gradient for a given wedge space $h(x)$ may be determined. In the following equations, only the pressure gradient in the wedge-shaped gap is derived, which is the basis for the notations in Fig. 4.10. The thickness of the fluid film at a given point equals:

$$h(x) = h_0 + \frac{l-x}{l}(h_1 - h_0) \quad (4.21)$$

when $0 \leq x \leq l$.

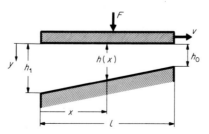

Fig. 4.10 Notations for converging fluid wedge

For further mathematical consideration it is useful to substitute:

$$m' = \frac{h_1}{h_0} - 1 \qquad (4.22)$$

for the wedge constant, and to replace the wedge-gap size at the point of maximum pressure h^* in equation (4.20) with:

$$h^* = k^* h_0 \qquad (4.23)$$

Substituting the equations (4.21), (4.22) and (4.23) into equations (4.20) we have, after some transposition:

$$dp = \frac{6\eta v}{h_0^2} \left[\left(1 + m'\frac{l-x}{l}\right)^{-2} - k^*\left(1 + m'\frac{l-x}{l}\right)^{-3} \right] dx \qquad (4.24)$$

After integration, this becomes:

$$p = \frac{6\eta v}{h_0^2} \frac{l}{m'} \left[\frac{1}{1 + m'\frac{l-x}{l}} - \frac{k^*}{2\left(1 + m'\frac{l-x}{l}\right)^2} + C \right] \qquad (4.25)$$

The constant of integration C as well as the constant k^* are governed by the boundary conditions:

$$x = 0 \rightarrow p = 0 \qquad (4.26)$$
$$x = l \rightarrow p = 0 \qquad (4.27)$$

Hence:

$$k^* = \frac{2m' + 2}{2 + m'} \qquad (4.28)$$

$$C = \frac{-1}{2 + m'} \qquad (4.29)$$

By substituting the values given in (4.28) and (4.29) into equation (4.25), the pressure function for the wedge gap is obtained, i.e.:

$$p(x) = \underbrace{\frac{6\eta v}{h_0^2} \frac{l}{m'} \left[\frac{1}{1 + m'\frac{l-x}{l}} - \frac{1}{2 + m'} - \frac{2m' + 2}{2(2 + m')\left(1 + m'\frac{l-x}{l}\right)^2} \right]}_{= K_p} \qquad (4.30)$$

To give a neater presentation, the last term in equation (4.30) is replaced by K_p, giving:

$$p(x) = \frac{6\eta v l}{h_0^2} K_p(x, m') \qquad (4.31)$$

Table 4.2 shows a number of values for the factor K_p for different wedge constants m' and various positions in the fluid wedge x/l.[34] It may be noted

Table 4.2 Values for the factor K_p in equation (4.31)

$\frac{x}{l}$	$m' = 0.6$	$m' = 0.8$	$m' = 1.0$	$m' = 1.2$	$m' = 1.4$	$m' = 1.6$	$m' = 1.8$	$m' = 2.0$
1.0	0	0	0	0	0	0	0	0
0.9	0.0185	0.0220	0.0248	0.0269	0.0285	0.0297	0.0306	0.0312
0.8	0.0294	0.0340	0.0370	0.0390	0.0402	0.0408	0.0410	0.0408
0.7	0.0348	0.0390	|0.0414|	|0.0426|	|0.0429|	|0.0426|	|0.0419|	|0.0410|
0.6	|0.0360|	|0.0393|	0.0408	0.0411	0.0406	0.0396	0.0384	0.0370
0.5	0.0341	0.0364	0.0370	0.0366	0.0356	0.0343	0.0328	0.0312
0.4	0.0299	0.0313	0.0312	0.0304	0.0292	0.0278	0.0263	0.0238
0.3	0.0240	0.0246	0.0242	0.0233	0.0220	0.0208	0.0187	0.0182
0.2	0.0168	0.0170	0.0165	0.0156	0.0147	0.0137	0.0127	0.0118
0.1	0.00876	0.00869	0.00831	0.00780	0.00725	0.00672	0.00591	0.00574
0	0	0	0	0	0	0	0	0

☐ = Maximum pressure.

that maximum pressure occurs at approximately $x = (0.6, \ldots, 0.7)l$, according to the wedge constant m'.

In practice, only the load-carrying capacity of the wedge surface is usually of interest, i.e. the mean effective hydrodynamic pressure p_m on the wedge surface area, which may be obtained by integrating equation (4.30) for the length of the wedge surface:

$$p_m = \frac{1}{l} \int_{x=0}^{x=l} p(x)\, dx \qquad (4.32)$$

After integrating and introducing a factor K_{pm} according to equation (4.30), we obtain:

$$p_m = \frac{6\eta v l}{h_0^2} K_{pm} \qquad (4.33)$$

The values for K_{pm} for various wedge constants are listed in Table 4.3.[34]

Table 4.3 Values for the factor K_{pm} in equation (4.33)

m'	K_{pm}
0.6	0.0235
0.7	0.0247
0.8	0.0255
0.9	0.0261
1.0	0.0265
1.2	0.0267
1.4	0.0265
1.5	0.0263
2.0	0.0246

It may be seen from Table 4.3 that the factor K_{pm} is not significantly affected by changes in the wedge constant, i.e. the load-carrying capacity of the fluid wedge is only marginally influenced by the angle of the wedge. For this reason, an approximate value for K_{pm} is perfectly adequate for all practical purposes, i.e. $K_{pm} = 0.025$.

The load-carrying capacity of the fluid wedge may therefore be stated to be (approximately):

$$F = p_m lb = 0.025 \frac{6\eta v l^2 b}{h_0^2} \qquad (4.34)$$

The considerations above have ignored the losses occurring at the sides of the fluid wedge. A series of experiments[35,36] have shown that allowance may be made for these by introducing a correction factor ψ in equation (4.33) so that:

$$p_m = \frac{6\eta v l}{h_0^2} K_{pm} \psi \qquad (4.35)$$

Table 4.4 gives values for this correction factor ψ dependent upon the ratio of the wedge breadth b to its length l for the two wedge constants $m' = 1$ and $m' = 2$.

Table 4.4 Correction factor ψ in equation (4.35) allowing for side losses

b/l	ψ when $m' = 1$	ψ when $m' = 2$
0.5	0.19	0.22
1.0	0.44	0.45
2.0	0.69	0.71
4.0	0.84	0.85
∞	1.00	1.00

The static stiffness k of the fluid wedge is obtained from

$$k = \frac{dF}{d(\Delta h)} \qquad (4.36)$$

where Δh is a load-dependent displacement. The load-carrying capacity of the fluid wedge is obtained from equations (4.34) and (4.35) as:

$$F = \frac{6\eta v l^2 b}{h_0^2} K_{pm} \psi$$

Any additional load ΔF results in a gap reduction Δh. Hence:

$$F + \Delta F = \frac{6\eta v l^2 b}{(h_0 - \Delta h)^2} K_{pm} \psi \qquad (4.37)$$

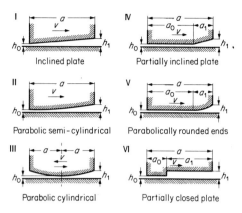

Fig. 4.11 Gap shapes for hydrodynamic pressure formation

By differentiating (4.37) we get:

$$\left[\frac{dF}{d(\Delta h)}\right]_F = 2\frac{6\eta v l^2 b}{(h_0 - \Delta h)^3} K_{pm} \psi \qquad (4.38)$$

Thus we may state the static stiffness for a given load F as:

$$k_F = \frac{2\sqrt{[(F + \Delta F)^3]}}{\sqrt{(6\eta v l^2 b K_{pm} \psi)}} \qquad (4.39)$$

(See also later in section 4.2.1.2).

The considerations so far have been with regard to a wedge-shaped fluid film. Figure 4.11 shows a number of other gap shapes in which hydrodynamic

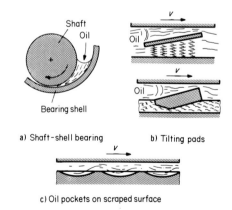

Fig. 4.12 Typical fluid wedges in machine-tool construction: (a) shaft-shell bearing, (b) tilting pad, (c) oil pockets

pressures may be formed. The pressure gradients for these shapes may be calculated precisely using equation (4.20), providing the function $h(x)$ is known.

Hydrodynamic bearings and guideways in machine tools frequently employ the fluid-wedge principle, as may be seen from the sketches in Fig. 4.12. In plain cylindrical bearings, the fluid wedge is formed by the eccentric positioning of the shaft in relation to the bearing bore or shell. Tilting-pad elements are most frequently used in hydrostatic thrust bearings, but can also be found in radial bearings. The pads are either spring loaded or their movement is merely due to the elasticity of the material into which they are embedded.

In the case of hydrodynamic slideways, microfluid wedges are formed in the imperfections of the sliding surface. It follows that very good surface finishes do not always provide the best conditions for hydrodynamic slides. Surface scraping provides oilpockets, which assist in the formation of hydrodynamic lubrication fluid films.

4.1.1.3 Types of friction

When two adjacent bodies move in relation to each other (as, for example, in the case of a slide on a machine bed or a shaft in its bearing), then various forms of friction may occur, governed by the speed and the medium between the surfaces,[37] as illustrated in Fig. 4.13. In order to set a stationary body into motion, the 'static friction' must be overcome. In order to maintain the motion, the moving friction must be surmounted. Moving friction is subdivided into sliding friction, rolling friction and rolling/sliding friction.

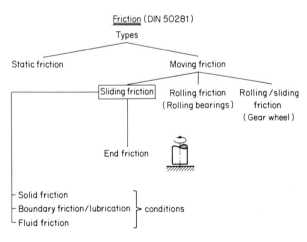

Fig. 4.13   Types of friction[37]

Sliding friction is the type of friction occurring in plain bearings and slideways. The following subdivision may be made:

(a) Solid friction (or dry friction, Coulomb friction), i.e. there is no lubricating film between the moving parts.
(b) Boundary friction (boundary lubrication). Although a lubricating film is present in the contact zone, the load-carrying capacity is inadequate for the complete separation of the parts moving relative to each other, so that contact between them remains (solid and fluid friction).
(c) Fluid friction. A continuous layer of a lubricating medium (e.g. oil, air, graphite) exists between the moving parts.

Rolling friction occurs by the rotation of circular, symmetric bodies, due to the elastic deformation within the contact area (rolling bearings, wheel on rail).

Rolling/sliding friction is present when flanks of gear teeth are in mesh, where in addition to the rolling friction a sliding action occurs governed by the kinematic conditions.

4.1.1.4 The Stribeck curve

In hydrodynamic slideways and bearings, the frictional conditions and the friction force F_R, as well as the coefficient of friction μ, vary in accordance with the sliding velocity:

$$\mu = \frac{F_R}{F_N} \qquad (4.40)$$

where F_N is the normal force component.

This relationship is presented in the so-called Stribeck curve, shown in Fig. 4.14, where the coefficient of friction μ, which determines the friction force F_R for a given normal force F_N, is plotted against the sliding velocity v. Different frictional condition stages are indicated on the sliding velocity axis of the

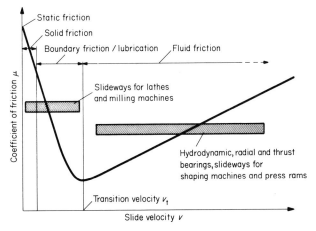

Fig. 4.14 Stribeck curve

Stribeck curve (see section 4.1.1.3). The so-called 'static friction' occurs between two stationary guide elements which are in contact with each other without being separated by a lubricating film. At very low sliding velocities, there is still no full load-carrying lubricating film; hence solid friction occurs in this velocity range, which is normally associated with high wear rates of the sliding surfaces. As the sliding velocity increases, the hydrodynamic load-carrying component increases. The solid-body contact component correspondingly declines (boundary friction/lubrication) until, at a particular sliding velocity (transition velocity v_t), the whole of the load is supported by the lubricating film (fluid friction). At this stage, the sliding surfaces are fully separated from each other by the lubrication film, so that no wear takes place. Hence it is possible to maintain movement at a velocity greater than the transition velocity.

Guideways on cutting and erosion machine tools are normally operated in the boundary lubrication range as the sliding velocities (feed velocities) are relatively low. Higher sliding velocities occur in machine elements which actually perform the work motions (e.g. shaping machine and press rams) as in such cases their speed of movement is governed by the relatively high cutting speeds required. Consequently, the slideways of such machines can operate in the fluid friction range of the Stribeck curve. Similarly, hydrodynamic thrust and radial bearings on main spindles usually run under fluid friction conditions.

Comparisons between the friction characteristics (Stribeck curves) of hydrodynamic guides with hydrostatic and roller guideways is given in Fig. 4.15. For a given oil viscosity, pressure and separation gap, the slope of the friction characteristic curve is constant in the fluid friction range for hydrostatic and hydrodynamic guides and bearings.

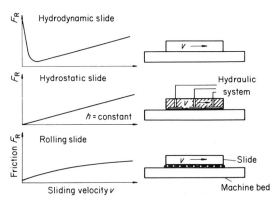

Fig. 4.15 Friction characteristics of different slideways

4.1.1.5 Stick-slip effect
When low sliding velocities in the boundary lubrication range are operating, irregular movements frequently occur which are a result of the stick-slip

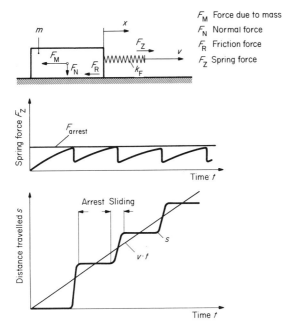

Fig. 4.16 Diagram of the stick-slip effect

phenomenon. This effect, which may be related to the negative slope of the Stribeck curve in the boundary lubrication range, manifests itself by a periodic stop–start motion.[38] Figure 4.16 illustrates this in a diagrammatic form. The tensile force F_Z which is applied by means of a spring fitted to the slide and the distance travelled s are shown as a function of time t. The spring, with its stiffness k_F, represents the elastic components of the drive (e.g. the stiffness of the feed shaft, feed gear, etc.). In order to set the slide into motion, the static friction must first be overcome, i.e. the spring is extended until the spring force equals the static friction force. The slide will then start to move, and between the slide and slideway there will remain only a fairly low frictional force, as may be seen from the Stribeck curve, resulting in an acceleration of the movement (jolt). This causes the tension on the spring to be reduced, so that movement of the slide stops. The static friction must now be overcome again, and this starts a new cycle.

The theoretical treatment of the stick-slip effect is by application of the differential equation of the slide motion:

$$m\ddot{x} + F_R + k_F x = k_F v t \qquad (4.41)$$

This can be solved with the forces acting on the slide in equilibrium:

$$F_Z = F_R = F_M \qquad (4.42)$$

where F_Z = spring force
       F_R = frictional force
       F_M = force due to mass

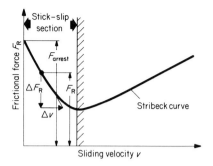

Fig. 4.17 Stick-slip section in the Stribeck curve

Ignoring the damping effect in the drive elements, the system is damped solely by the frictional force F_R, which is indicated in the first part of the curve in Fig. 4.17:

$$F_R \approx F_{\text{Arrest}} + C_1 v \qquad (4.43)$$

where $C_1 = \Delta F_R / \Delta \dot{x}$, corresponding to the slope of the Stribeck curve.

In the boundary range, the slope of the Stribeck curve C_1 is negative; hence the damping of the system reduces when the sliding velocity v increases. Therefore, the equation for the motion in the boundary friction range may be stated as:

$$m\ddot{x} - C_1 \dot{x} + k_F x = k_F v t - F_{\text{Arrest}} \qquad (4.44)$$

The tendency towards stick-slip effects may be reduced by:

(a) reducing the drop in friction value in the first section of the Stribeck curve;
(b) reducing the transition velocity by the use of a higher viscosity oil and/or larger supporting surfaces;
(c) increasing the static stiffness of the feed-drive components;
(d) reducing the masses.

A reduction in the drop in friction value can be obtained by the use of high polymerized additives in the lubricating oil or by plastic coatings on the guideway elements (see section 4.1.2.1), as shown in Fig. 4.18.

4.1.1.6 Raw material pairings and wear factors

Slideways on machine tools frequently operate in the boundary lubrication range. For this reason, the guiding elements are subject to potential wear. Whilst it may be possible to correct even wear along the whole length of the guideway by adjustments (e.g. gib strips), uneven wear requires re-machining.

For example, if on a centre lathe the guideways are mostly worn near the spindle as a result of the machine being used mainly for chucking work, as

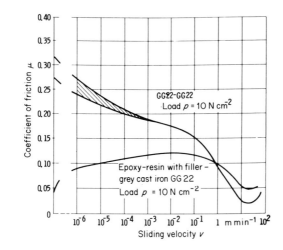

Fig. 4.18 Friction characteristics of coated and non-coated guideways (Gleitbelagtechnik GmbH)

shown in Fig. 4.19, then the turning of long components to close tolerances is no longer possible. The wear rate of a plain bearing or slideway is strongly influenced by:

(a) the surface preparation of the sliding surfaces;
(b) the raw material pairing;
(c) the pressure per unit area.

Influence of the preparation of the sliding surface. During its running-in period, the wear rate of a slide for a given sliding distance is governed very much by the method of its preparation. The greater the 'peak-to-valley' surface roughness is on a new slide, the more rapid will be the wear during the running-in period. The experimental data depicted in Fig. 4.20 show the wear rates of a bronze component made from Sn Bz 8 for the sliding distances when in contact with grey cast iron (GG25) surfaces having differing finishes. After a certain sliding distance the wear rate becomes constant because the roughness peaks have been removed.

Influence of raw material pairing. The influence of the pairing of the raw materials for sliding components on the frictional characteristics, and hence on the wear rate, is very high for solid bodies operating in the boundary lubrication range. Figure 4.21 evaluates the wear rates for different material pairings between 20 and 35 km sliding distance, so that the running-in phase is not a factor.

One of the sliding surfaces is made from grey cast iron GG25, whilst the other is as stated in the diagram. It can be seen that when the experimental

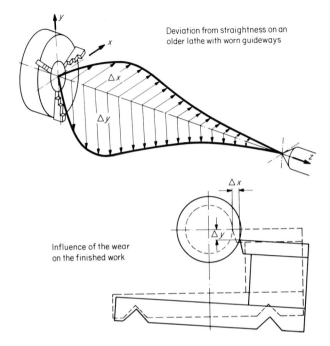

Fig. 4.19 Uneven wear, on the guideways of a centre lathe

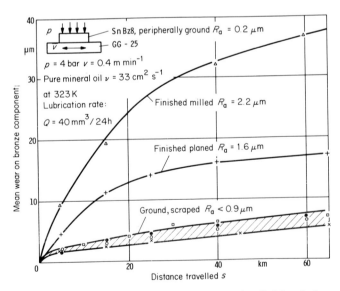

Fig. 4.20 Frictional wear in relation to the finish of the mating surface

109

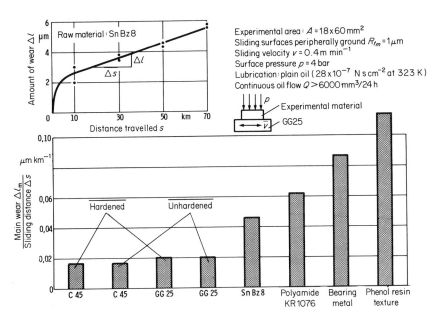

Fig. 4.21 Frictional wear rates of various raw material pairings

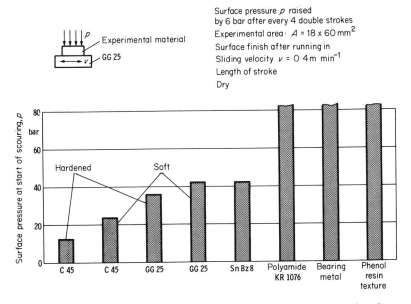

Fig. 4.22 Influence of raw material pairings upon the tendency of surface scouring

materials are steel (C45 hardened and C45) or cast iron (GG25 hardened and GG25) the wear rates are relatively low, whilst when the bronze-facing bearing metal and plastic materials are used, considerable increases are indicated.

The wear rate is not the only consideration when choosing a raw material pairing, as attention must also be paid to the tendencies of so-called 'scouring', when the lubricant is squeezed out or supplied in inadequate quantities.

In Fig. 4.22, the experimental results are given for tests in which the surface pressure was steadily raised on a flat plate made from grey cast iron GG25 having a sliding velocity of $v = 0.4$ m min^{-1}. All other factors such as surface finish, length of stroke, test area, etc., were kept constant. Whilst scouring occurs on the test piece at relatively low surface pressures when mating with steel, cast iron and bronze facing, even surface pressures in excess of 800 N cm^{-2} do not cause scouring when the plastic materials and the bearing metal are in contact. Such materials have therefore good 'emergency' running conditions.

Influence of surface pressure and positioning of the guideway surfaces. The wear rate increases approximately in direct proportion to the surface pressure. Consequently, the width of guideways should be of adequate proportions in relation to the direction of the applied force. Figure 4.23 shows, in diagrammatic form, how a constant force F leads to varying surface pressures (p_A and p_B) and hence to differing wear values (δ_A and δ_B) on the guiding surfaces A and B, when applied in a variety of directions.[39] The wear values δ_A and δ_B cause a displacement of the slide being guided.

Two conclusions may be reached from Fig. 4.23 which affect design considerations:

(a) The guideway surfaces should be, as near as possible, normal to the resultant applied force.

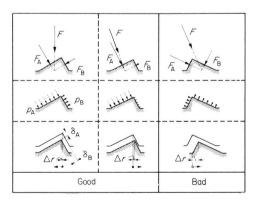

Fig. 4.23 Influence of the load on surface pressure, wear and work accuracy (Saljé[39])

(b) The wear caused should, as far as possible, initiate only a tangential relative displacement of the tool and work at the machining point. This will minimize any inaccuracies resulting from wear on the guideways.

4.1.2 Hydrodynamic slideways

Hydrodynamically lubricated linear guides are the most common types used in machine-tool construction in a variety of designs. This is due to their good damping properties, as well as the high degree of accuracy and rigidity attainable at relatively low design and production costs.

4.1.2.1 Guideway elements and design features

The basic geometric shapes of the guiding elements are derived from rectangles, triangles and circles, as shown in Fig. 4.24.[39] In order to be able to remove play in linear guides, gib strips are usually fitted. The fundamental shapes illustrated are the basic forms for guide designs used in practice. The most common guide shape used today is the flat guide, which is easily manufactured and provides good rigidity. For trouble-free side control of a table, narrow guides should be used as shown in the lower part of Fig. 4.25. The distance b on the guide surface should be kept small ('narrow' guide) in order to minimize 'cross winding' ('sash window' or 'drawer' effect) and to reduce the effect of thermal influences on the clearance. Figure 4.25 clearly shows that the slide with a wide distance between the side-guiding surfaces is subjected to greater play through heating effects (e.g. hot swarf lying on the table) than the slide which is designed with the narrow guide principle.

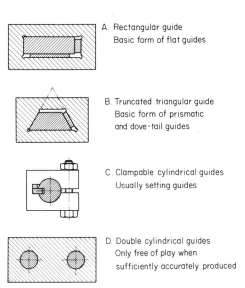

Fig. 4.24 Basic shapes of guides and guideways (Saljé[39])

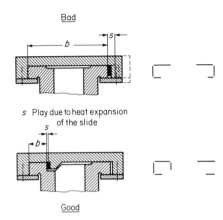

Fig. 4.25 Flat guides with, and without the application of the narrow guide principle

Figure 4.26 illustrates some examples of designs utilizing the narrow-guide principle. Example A has a complicated retaining strip which is difficult to fit. Moreover, the internal guide face of the narrow guide on the bed is difficult to produce, owing to its awkward position. Conversely, the narrow guides on the beds shown on the B and C designs are more easily machined as they are exposed on both sides. In the case of design D, the narrow guide is in the centre of the table. This prevents the bed being weakened by the narrow guide with regard to the vertical guide faces, and a symmetrical expansion of both sides of the table may be expected as a result of any heating effects.

Longitudinal guides which are designed on a triangular basis (see Fig. 4.24) control the table in two directions, making additional narrow guides superfluous. On lathes, vee guides in combination with flat guideways are frequently

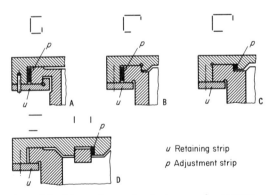

Fig. 4.26 Examples of designs of narrow guideways

113

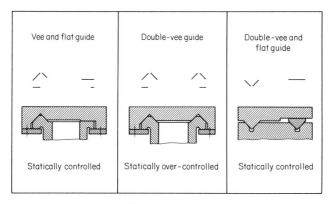

Fig. 4.27 Vee/flat and double-vee guideways

employed, and in much rarer cases, double-vee guides are used (see Fig. 4.27). A double-vee guide provides a first clas statically and dynamically combined control. The disadvantages are, however, the static over-control and the high manufacturing costs due to the more difficult fitting work. Thermal expansions of the slide at right angles to the direction of movement will tend to raise the table, which may cause binding. This guide principle is, therefore, mostly used for low-load precision manufacturing and inspection machines. On the other hand, vee and flat guides are statically controlled. The table may, under the influence of a heat source, expand in the direction of the flat guide without binding. When compared with a simple flat guideway (Fig. 4.25) it can be seen that there is one less guiding face. Vee and flat guides exhibit the advantageous feature of self-cleaning (dust and sludge flow away).

Another type of guideway based upon the triangular shape is the 'dove-tail' guide, as shown in Fig. 4.28. Dove-tail guides have only four guide faces, and are therefore relatively small. Their manufacture is somewhat expensive, due

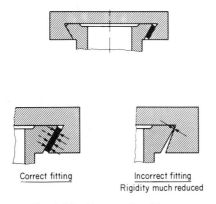

Fig. 4.28 Dove-tail guideway

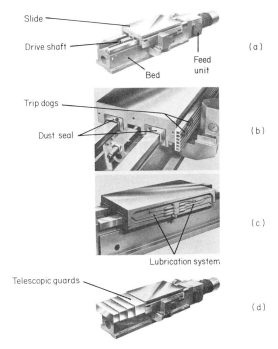

Fig. 4.29 Flat guides and accessories of a slide unit (DIAG–Honsberg)

to the careful fitting necessary. The accuracy of fitting has a great influence upon the rigidity of this type of guide.

The following pictures show examples of linear guides for machine-tool slides and saddles. In Fig. 4.29 a slide unit is pictured which is, for example, built into transfer machines. Hardened interchangeable guideways are bolted on to the machine bed. The lubrication points on the guides are fed with lubricant by connection to a central lubrication system (c). In order to protect the guideways from damage, dust seals (b) and telescopic guards (d) are fitted to the slide.

In Fig. 4.30, the arrangement of all the guideways for the slide and milling head motions of a universal milling machine may be seen. All guides on this machine are of the flat type.

Figure 4.31 illustrates the positions of the guides for the saddle and tailstock on a NC inclined bed lathe. The upper guide supports the saddle (working guide), whilst the lower guideway is for the tailstock (setting guide). The features which may be noted on this layout include the internal retaining plate, the good rigidity inherent in the design and the easily machined flat guides which are readily protected and which, in conjunction with the inclined bed, allow easy swarf disposal.

Guideways on machine beds and columns are normally made out of ground

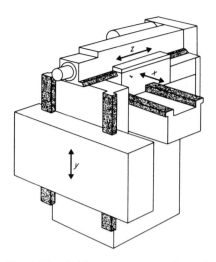

Fig. 4.30 Guideways on a universal milling machine (Deckel)

steel strips which are bolted or clamped on to the machine frame, as shown in Fig. 4.32. The advantage of this is that worn or damaged guideways are interchangeable. Moreover, the guideways do not have to be flame-hardened, which is a necessary procedure on guides which are solid with the main body of the machine (usually of cast iron) to provide adequate wear resistance. On long guideways several steel strips are fitted in line with each other. In some instances, instead of fitting steel guide strips, hardened and ground steel bands are bonded to the bed (see Fig. 4.33).

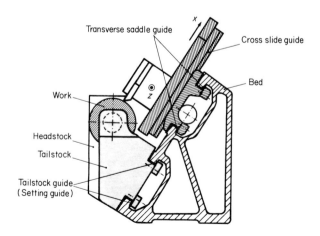

Fig. 4.31 Guideways on a NC inclined bed lathe (Gildemeister, Max Müller)

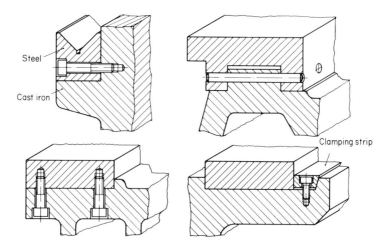

Fig. 4.32 Steel inserted guideways (Schaerer, Giddings and Lewis, Scharmann)

In recent designs, plastic coatings are applied to guide surfaces in ever-increasing instances; these are used in combination with ground-steel guideways. Such a material pairing results in excellent emergency running conditions and a favourable relationship in the boundary lubrication range of the Stribeck curve, virtually eliminating all tendencies towards the stick-slip effect (see sections 4.1.1.4 and 4.1.1.5).

Plastic coatings are applied to guideways either by adhesive bonding or by the use of moulding techniques. Figure 4.34 shows three stages of the mould-

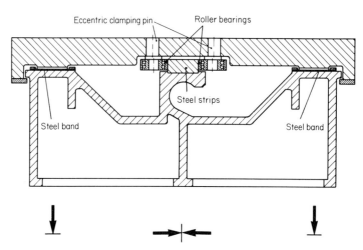

Fig. 4.33 Table guides utilizing tensioned steel bands (Scharmann)

Fig. 4.34 Construction of a plastic faced guideway
(Waldrich/Coburg)

ing process. The guideways of the machine are given a ground finish, whilst the contact surfaces of the slide are prepared with a comparatively rough surface (milled or planed). The upper picture shows the prepared guide faces on the underside of the table, which are to be smoothed with the plastic material. After the table guideways are covered with the plastic, the table is laid into the corresponding guides on the bed, the guides having previously been covered with a releasing agent to prevent the plastic material adhering. Distance pieces are placed between the table and the bed to give the correct form to the table guides and to ensure an even thickness of the plastic coating. The weight of the table and, if necessary, additional weights or bolts squeeze the surplus plastic out of the gap. The material remaining between the table and the bed is then allowed to harden.

4.1.2.2 Clamping devices
A large number of guides on machine tools are used as either working guides or setting guides, depending upon the particular work to be machined. Such guides must therefore be able to be clamped when stationary using a special clamping device.

Figure 4.35 illustrates a clamping device for a double-cylindrical guide of a drilling and milling machining centre. The slide is guided by four adjustable bushes. The guides are clamped by pressing clamping bars against the cylindrical pillars. The clamping forces acting when the clamps are applied are

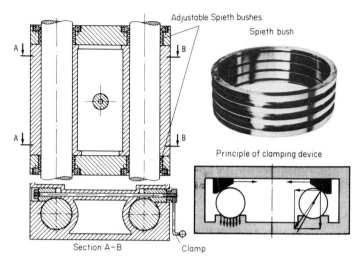

Fig. 4.35 Clamping device for a double cylindrical guide (Scharmann)

shown in the diagram. It should be noted that this guiding system, when accurately produced, provides close control characteristics. Owing to the static overlocation, any appreciable temperature differentials between the slide and the cylindrical guides will result in excessive loads on the guiding sleeves. The overall rigidity of such designs is normally governed by the elastic characteristics of the cylindrical columns.

A different clamping device for a flat guideway is shown in Fig. 4.36. The clamping function is activated by hollow rubber elements fitted into the retaining plate of the flat guide, internally pressurized by a hydraulic oil flow

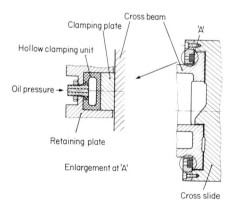

Fig. 4.36 Hydraulic clamping unit of the cross slide on a turret lathe (Schiess–Foriep)

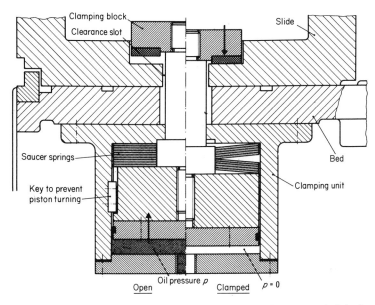

Fig. 4.37 Clamping unit for large guideways: left is open and right is clamped (Scharmann)

and exerting pressure on to the clamping unit against the cross beam guides. By the use of hollow rubber elements, the need for expensive hydraulic pistons and their associated sealing problems are obviated.

Figure 4.37 illustrates a section through a clamping unit which is suitable for the application of the larger clamping forces, necessary on heavy machine tools. A number of such cup-shaped clamping units are bolted on to the underside of the guideways of the bed. The slide is provided with a clearance slot, through which the clamping rods are fitted. Clamping is activated by a pack of saucer springs (right in the diagram). The clamp is released by means of hydraulic pressure (left). An advantage of such a clamping device is that the clamping force is unaffected by any failure of the hydraulic system.

4.1.2.3 Compensation for guiding errors

Guiding errors cause deviations from the planned relative motion between the tool and work. Errors are only fully transmitted into the accuracy of the work being produced when the deviations at the cutting point act normal to the surface of the material being cut. Deviations which are tangential to the machined surface have only a limited effect upon the accuracy of the work.

Errors on the work being machined which are traceable to faults in the guideways may be caused by:

(a) inaccurate production of the guideways;
(b) wear of the guideways;

(c) static deformations caused by the weight of the machine-tool components and/or by the cutting forces;
(d) thermal deformations caused by temperature differentials.

Accurately produced guideways are a prerequisite for precision machine tools. In order to ensure that the required accuracy can be produced, the expected static deformations—e.g. due to the weight of the slides—are compensated by appropriate corrective action during the manufacture of the slides. For example, cross beams for gantry machines are clamped on to the table of the grinding machine during the machining of their guideways in such a way as to allow for their subsequent static deformations.

The following preventive action may be taken to avoid work inaccuracies due to static deformations from cutting forces and mass effects:

(a) the most rigid construction possible;
(b) crowned scraping or grinding of the guideways;
(c) feed-back control devices (see Volume 3);
(d) balance weights, e.g. on the cross beam of large gantry milling machines;
(e) geometric adaptive control (see Volume 3).

4.1.2.4 Static and dynamic behaviour
The static and dynamic characteristics of a plain slide have a major influence on the characteristics of the whole machine tool. In this respect, the rigidity and damping effects in the contact zones are of prime importance.[39] The damping effect in the contact zone may be traced back to three basic causes:[38]

(a) material damping effects caused by hysteresis losses in the material due to elastic and plastic deformations of the surface finish peaks (normal action);
(b) friction damping caused by micro-movements in the boundary zones of the surface during vibrations (normal and tangential action);
(c) fluid damping (squeeze–film effect) during the 'squeezing-out' and 'drawing-in' action of the lubricating fluid (normal action).

In the case of smooth, well-lubricated sliding surfaces, it may be assumed that the fluid and friction damping effects overshadow the material damping effect (see Volume 4).

It is possible in the boundary lubrication zone for the contact rigidity of a hydrodynamic plain slide to be related to various components of the support load. In the boundary lubrication zone the guide faces are supported by contact with the solid body of the frame, as well as with the hydrodynamic lubrication film (see section 4.1.1.2).

Under non-stable conditions, i.e. when there are changes in the operating conditions (e.g. load, velocity), a transitory procedure occurs during which the conditions in the contact zone, e.g. the lubrication film thickness, are adapted to the new conditions. The variation of the thickness of the lubrication film caused by changes in load or velocity varies the volume of the lubrication medium between the sliding surfaces.[38] The lubricating fluid is thus squeezed out of the gap against the flow resistance or, conversely, is drawn in between the sliding surfaces when the load is released. The result of the displacement of the lubricating film (squeeze-film effect) is to produce a further force component normal to the sliding surface, known as the 'displacement force'. Hence the supporting force F_S in the boundary lubrication range consists of:

$$F_S = F_{Contact} + F_{Hydrodyn} + F_{Displacement} \qquad (4.45)$$

In relation to the changing gap size and support conditions caused by varying sliding velocities, the damping effect in the contact zone also varies.

The damping effect normal to the guide surfaces is governed by the geometric characteristics of the guides, the sliding velocity and the lubricating medium in use. The damping of the vibrations which occur tangential to the guiding surfaces is reduced as the sliding velocity increases, due to a reduced coefficient of friction[40] (see section 4.1.1.4).

4.1.3 Hydrodynamic plain circular bearings

Hydrodynamic main spindle-bearing systems are found in machine tools where the running of the bearing in the fluid friction range is an advantage

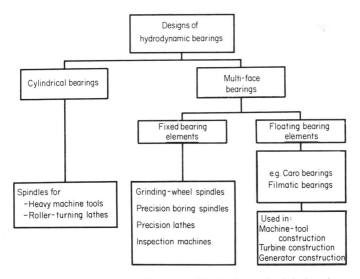

Fig. 4.38 Types and applications of hydrodynamic plain bearings for machine tools

and where the spindle is not frequently stopped and restarted. For slow-running spindles, hydrodynamic plain bearings are only rarely used as the wear rate is too great when running under boundary lubrication conditions.

According to their design, hydrodynamic plain bearings are classified into cylindrical and multi-face bearings, as shown in Fig. 4.38. Cylindrical hydrodynamic bearings are used in machine tools, e.g. on spindles of heavy roller-turning lathes. Multi-face bearings centralize the shaft more accurately, due to their wedge grip effect. Consequently, these are used in large quantities for machine tools. A further subdivision may be made in such bearings between those with fixed bearing elements and those having floating, self-aligning bearing pads.[41,42]

Bearings with self-aligning bearing elements offer the advantage of rapid adaptation to changing running conditions. Their main application is in turbine and generator construction, but as the self-aligning mechanism has a poor rigidity their use in machine tools is much rarer.

For machine tools such as grinding and precision boring machines, as well as precision lathes and inspection machines, multi-element bearings with fixed bearing pads are frequently used because of their simple construction.[42]

Before discussing in detail the various designs of hydrodynamic plain bearings, the pressure build-up and the starting characteristics are explained. For a detailed mathematical analysis of hydrodynamic plain bearings, the reader is recommended to study the VDI 2201 and VDI 2204.[43,44]

4.1.3.1 Pressure build-up and acceleration characteristics

The conditions described in section 4.1.1.2 are also valid for the pressure build-up in the wedge space of a hydrodynamic plain bearing. Figure 4.39 illustrates the pressure gradient of the lubrication film which occurs in a circular plain bearing, due to the eccentric position of the shaft in relation to

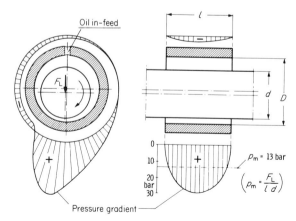

Fig. 4.39 Shaft displacement and pressure gradient of the lubrication film in a circular plain bearing

the bearing housing. Maximum pressure occurs just in front of the narrowest part of the gap, which is on the line joining the centres of the shaft and bearing (see Fig. 4.8). Behind the narrowest part of the wedge, a suction action produces a negative pressure (vacuum).[45,46] The pressure drops in the direction parallel to the axis of the shaft, side losses of the lubricant. The oil holes in circular plain bearings must be so arranged as not to interfere with the pressure build-up, which is governed by the loading conditions and direction of rotation.

The centre of the shaft lies eccentrically in relation to the bearing, but approaches alignment with it as the speed of rotation increases. This acceleration procedure is presented in the diagrams of Fig. 4.40.[46] A stationary shaft lies in direct contact with the bearing. The load is supported by direct solid contact. When the shaft starts to rotate, the static friction must first be overcome. Until this happens, the shaft rolls upwards along the wall of the bearing and then drops down again. In this starting phase, the shaft runs in an unstable manner.

After the hydrodynamic lubricating film is established, the centre of the shaft is displaced towards the centre of the bearing as the speed increases in accordance with the so-called Gümbels semi-circle.[46] In Fig. 4.40, the acceleration characteristics of a guided hydrostatic bearing under load F are shown

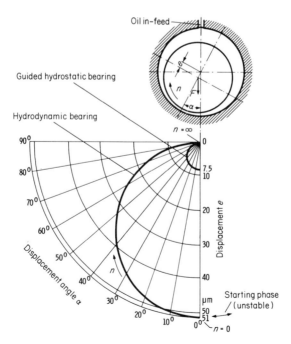

Fig. 4.40 Speed-dependent centre displacement in a cylindrical plain bearing (Gumbel's semi-circle)

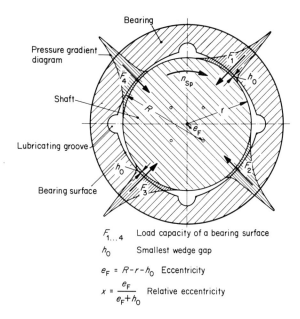

$F_{1...4}$ Load capacity of a bearing surface
h_0 Smallest wedge gap
$e_F = R - r - h_0$ Eccentricity
$x = \dfrac{e_F}{e_F + h_0}$ Relative eccentricity

Fig. 4.41   Diagram and fluid-film pressure gradients of a multi-face bearing

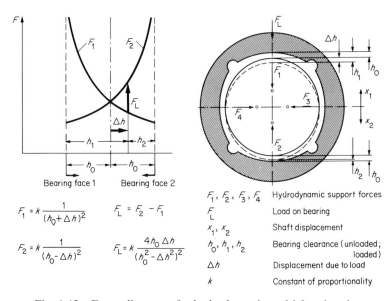

$F_1 = k \dfrac{1}{(h_0 + \Delta h)^2}$

$F_2 = k \dfrac{1}{(h_0 - \Delta h)^2}$

$F_L = F_2 - F_1$

$F_L = k \dfrac{4 h_0 \Delta h}{(h_0^2 - \Delta h^2)^2}$

F_1, F_2, F_3, F_4 Hydrodynamic support forces
F_L Load on bearing
x_1, x_2 Shaft displacement
h_0, h_1, h_2 Bearing clearance (unloaded; loaded)
Δh Displacement due to load
k Constant of proportionality

Fig. 4.42   Force diagram of a hydrodynamic multi-face bearing

for comparison, in which a hydrodynamic support acts in addition to the static oil pressure (see section 4.2).

In the case of multi-face bearings, a number of pressure gradients may be present which depend upon the pressure development; these will tend to centralize the shaft in its bearing as shown in Fig. 4.41. Apart from this advantage, the multi-face bearing runs more steadily and is very suitable when the direction of loading and/or the direction of rotation is varied.

Figure 4.42 shows the force diagram of a multi-face bearing. In the unloaded condition, the shaft is centralized by forces F_1 to F_4. In the force diagram, the unloaded condition is represented by the intersection of the two curves. When a load F_L is applied on the centralized shaft, the displacement is Δh. The effect of the support by four bearing faces is to markedly improve the rigidity $F/\Delta h$.

4.1.3.2 Design variations
In the following paragraphs, a number of specialized designs of hydrodynamic bearings are introduced which have been found in practice to be of particular value.

Figure 4.43 shows the so-called three-face bearing by Mackensen. The triangular profile of the bearing surface is produced by elastic deformation of the bearing, the outside of which is tapered and has a triangular section. As the bearing is pressed into its retaining bore, the triangular profile is formed on the bearing's internal diameter. The clearance between the bearing and its shaft can be set and adjusted by the amount the bearing is drawn into the bore in which it is fitted. Owing to the reduced contact area between the bearing and the fitting bore, the rigidity in this case is not very great.

Figure 4.44 shows a hydrodynamic plain bearing which is self-adjusting for temperature variations, consequential shaft distortions and misalignments.

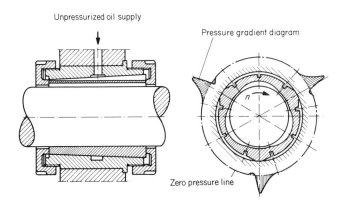

Fig. 4.43   Three-face bearing (Mackensen)

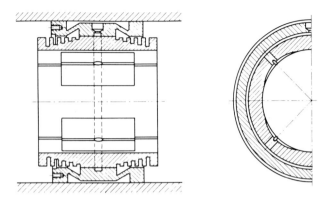

Fig. 4.44 Caro-expansion bearing

This so-called 'Caro-expansion bearing' consists of an externally ribbed bearing shell, from which any frictional heat generation within is readily abstracted by means of oil circulating around the fins. By means of the elastic suspension in the outer bearing shell and support by the ribs, any 'corner rubbing' due to shaft bending or misalignment is avoided. Similarly to the Mackensen bearing, this design, due to its resilient components, offers very poor rigidity when compared, for example, with the compact multi-face bearing shown in Fig. 4.41.

The aim of the so-called 'filmatic bearing' shown in Fig. 4.45 is to provide

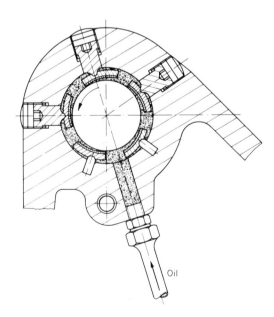

Fig. 4.45 Filmatic bearing

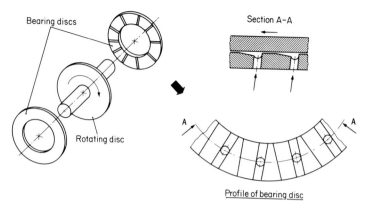

Fig. 4.46 Hydrodynamic thrust bearing

good centralization of the shaft by means of an optimum, self-adjusting lubricating wedge formation with varying rotational speeds. This bearing contains moveable bearing pads (see Fig. 4.12) which may be adjusted by means of setting screws. The clearance gap size h_0, i.e. the positioning of the bearing pads relative to the shaft, adjusts automatically with relation to the rotational speed. Thus the increased clearance provided at high speeds reduces the power loss. For trouble-free functioning of this bearing, it is necessary to entirely submerge the bearing pads in oil, in order to draw the bearing into the wedge space right at the beginning of rotation.

For the absorption of axial loads, hydrodynamic thrust bearings are used. Lubrication wedges are usually built into the bearing discs of plain thrust bearings to support the hydrodynamic pressure, as shown in Fig. 4.46. For trouble-free functioning of plain thrust bearings, it is necessary to ensure that no uneven pressures are formed on the faces of the bearing discs, even when the shaft is under a bending deformation. When the lubrication clearances are relatively small, the rigidity of the shaft will be correspondingly good.

4.1.3.3 Hydrodynamic spindle bearing units in machine tools
As already mentioned in the beginning of this chapter, in machine-tool construction hydrodynamic plain bearings are mainly used for precision machines. A particularly useful application is for grinding-machine spindle bearings, where running conditions are almost ideal, i.e. the rotational speeds are sufficiently high and nearly constant. A grinding-wheel spindle does not have to be frequently stopped and started, as it continues to rotate even when the component is reloaded.

Similar operating conditions also apply to bearings for precision boring-machine spindles. Figure 4.47 shows a section through a hydrodynamic bearing unit for such a machine, incorporating a multi-face bearing as outlined in Fig. 4.41. This unit is notable for its very simple design. The bearing housings are made oversize and then shrunk into the retaining bores in the spindle

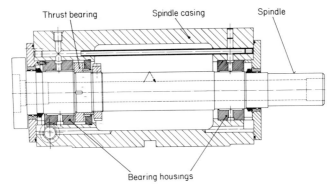

Fig. 4.47 Multi-face bearings for a precision boring machine spindle

casing. The compact construction in combination with the multi-face bearings results in a very good rigidity.

Figure 4.48 shows the general arrangement of a grinding-wheel spindle bearing with hydrodynamic multi-face bearings incorporating resilient bearing shoe supports. The tapered bearing faces of the two plain bearings obviate the need for an additional thrust bearing. The arrangement is that found in taper roller bearings, in 'O' layout (see section 4.3). The bearing shoe sup-

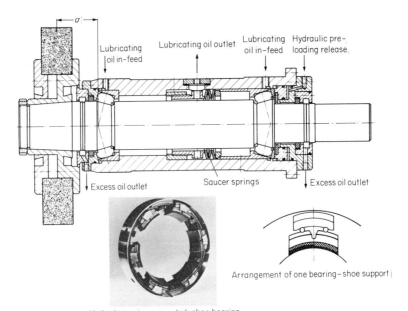

Fig. 4.48 Grinding-wheel spindle bearing unit with hydrodynamic, supported bearing shoes (FAG–Kugelfischer)

ports are made with spherical backs which provide a self-adjustment. The wedge-shaped lubrication gap is formed by elastic deformation of the lower part of the bearing shoe supports. On the drive end of the spindle the bearing is axially adjustable, whilst on the working end it is fixed. Consequently, any expansion due to a temperature rise of the spindle has no effect on the grinding accuracy. The two bearings are pressed against the running faces of the spindle by means of a set of saucer springs. During warming up and cooling down of the spindle, the pre-loading is hydraulically released to ensure that there is adequate clearance and lubrication space during these critical periods. Thus the starting wear is minimized. When the operating speed is reached the oil pressure is removed and the bearings are pre-loaded by the saucer springs. The lubrication film is then only a few micrometres thick, depending upon the viscosity of the lubricating oil.

4.2 Hydrostatic plain bearings and guideways

In the case of hydrostatic bearings the hydrodynamic effect for the provision of the load-carrying pressure in the lubricating gap, as described in the sections above, is largely replaced by the hydrostatic principle, in which the oil pressure is generated externally to the bearing. Consequently, the contact faces of two machine components which slide against each other are separated during operation by a permanent oil film.[47] The provision and maintenance of the volume of oil necessary for this lubricating film is provided by an oil-supply system which is separate from the bearing. In contrast to the ordinary plain bearing, the thickness of the oil film is therefore virtually independent of the sliding speed. Moreover, this type of lubrication ensures that the bearings are wear-free and without starting friction; they are running constantly under fluid lubrication conditions, which makes any tendency for stick-slip impossible, even when the sliding velocity is low.

4.2.1 Fundamentals and basic operating principles

The basic concept of a hydrostatic guide is shown in Fig. 4.49. In one of the two sliding surfaces recesses known as oil cells are provided. These are sur-

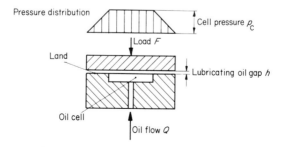

Fig. 4.49 Basic concept of a hydrostatic bearing

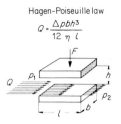

Fig. 4.50 Flow in a parallel gap

Hagen-Poiseuille law
$$Q = \frac{\Delta p b h^3}{12 \eta l}$$

Q Flow quantity
$\Delta p = p_1 - p_2$ Pressure drop over length
b Breadth of gap at right angles to flow
h Height of oil gap
η Dynamic viscosity
l Length of gap in direction of flow

rounded with 'lands' and supplied with oil. The distance between the land and the sliding surface above it is the lubricating oil gap h (approximately $20\ \mu\text{m} \leqslant h \leqslant 80\ \mu\text{m}$). The gap forms a hydraulic resistance, which impedes the oil flow from the oil cell and thus causes the pressure build-up which is the reaction to the external load. The difference between the pressure in the cell and atmospheric pressure is known as cell pressure p_C. This pressure falls to zero at the bearing land, as indicated in the upper diagram, because at the outer edges of the land usually only atmospheric pressure is acting. Figure 4.50 shows a section of the land surrounding the oil cell. The flow quantity in the parallel gap over the land may be calculated using the 'Hagen–Poiseuille' law for linear flow.

By analogy with Ohm's law in electrical technology and considering the pressure differential Δp equivalent to a voltage V and the flow quantity Q to a current I, the hydraulic flow resistance in the parallel gap may be stated as:

$$R_C = \frac{\Delta p}{Q} = \frac{12 \eta l}{b h^3} \qquad (4.46)$$

The electric circuit analogous to a bearing cell and the oil supply is shown on Fig. 4.51,

where R_C = hydraulic resistance of the bearing cell
            R_K = hydraulic resistance in the oil supply lines (capillary tube)
            p_C = cell pressure
            p_P = pump pressure

This analogy to Ohm's law and to the Kirchhoff network laws is of great value for the mathematical analysis of hydrostatic bearings, which, in general, consist of cells and resistances in the oil circuit.

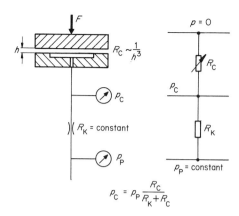

Fig. 4.51 Hydrostatic bearing without retaining plate and its electrical analogy

Figure 4.52 shows by a three-dimensional presentation the pressure distribution at the bearing cells and lands. The pressure gradient over the lands may be regarded as linear to a first approximation. Hence it may be assumed that the full cell pressure is acting beyond the cell up to the centre of the land.

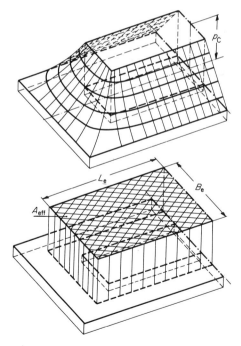

Fig. 4.52 Three dimensional representation of the pressure distribution of the effective area

This area, over which the full cell pressure is assumed to be acting, is known as the 'effective' area A_{eff}, and is shown shaded in the lower half of Fig. 4.52. It may be calculated by knowing the effective length L_e and the effective breadth B_e.

These relationships are shown in the diagrammatic presentation of a hydrostatic bearing cell in Fig. 4.53. The following relationships and definitions may be made:

Length of outflow surface l	Width of land in direction of flow
Breadth of outflow surface $b = 2(B_e + L_e)$	Periphery of cell at the centre-line of the land (perpendicular to the direction of flow)
Flow quantity $Q = \dfrac{p_C b h^3}{12\eta l}$	Hagen–Poiseuille law
Load $F = A_{\text{eff}} p_C$	Force equilibrium
Land area $A_R = lb$	Area which is critical for friction
Hydraulic resistance at cell $R_C = \dfrac{p_C}{Q} = \dfrac{12\eta l}{bh^3}$	Based upon the Hagen–Poiseuille law

For different outflow lengths (e.g. l_1 along the effective bearing breadth B_e and l_2 along the effective length L_e, as shown on Fig. 4.54) the total flow quantity is subdivided into the flow quantity Q_1 over the breadth $2B_e$ and Q_2 flowing over the breadth $2L_e$, enabling the hydraulic resistance for the cell R_C to be calculated. The analogous circuit diagram with two resistances R_1 and R_2

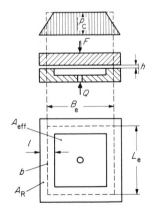

Fig. 4.53 Diagrammatic presentation of a hydrostatic bearing cell

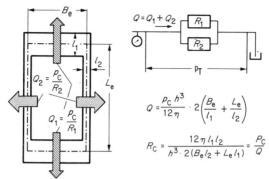

Fig. 4.54 Flow and hydraulic resistance at the cell for different lengths of outflow

in parallel is used to derive the following relationships:

$$R_1 = \frac{12\eta}{h^3} \frac{l_1}{2B_e}, \qquad R_2 = \frac{12\eta}{h^3} \frac{l_2}{2L_e} \qquad (4.47)$$

$$R_C = \frac{R_1 R_2}{R_1 + R_2}, \qquad R_C = \frac{12\eta l_1 l_2}{h^3 2(B_e l_2 + L_e l_1)} = \frac{p_C}{Q} \qquad (4.48)$$

4.2.1.1 Oil supply systems

In order to support eccentric loads hydrostatic bearings and guides are made with multiple-bearing cells. The supplies to the cells must be independent of each other, so that varying pressures may be built up in different cells in response to the changing equilibrium conditions.[47,48] Two possible oil supply systems are shown in Fig. 4.55. The system shown on the right, 'one pump per cell', has a high load-carrying capacity since the cell pressure is limited only by the maximum pump pressure available. However, the provision of many pumps involves high capital and running costs. The 'single pump with resistors' system is widely employed for economic reasons. In this system the compensating resistances are obtained by varying the cross-sections in the supply lines between the pump and cells.

Resistors which are built as capillary tubes are preferred because the flow is then governed by the viscosity, as is the cell resistance itself. The Ohm's law analogy is applied for the design of the oil supply system to determine the capillary tube resistances to suit the cell resistances and the flow quantity.

The flow quantity through a capillary tube resistor and the resistance are given by:

$$Q = \frac{\Delta p \pi r_K^4}{8\eta l_K} \qquad \text{or} \qquad R_K = \frac{8\eta l_K}{\pi r_K^4} \qquad (4.49)$$

where Δp = pressure drop over the capillary tube length
r_K = radius of capillary tube
l_K = length of capillary tube

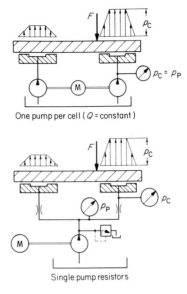

Fig. 4.55 Different oil supply systems for a hydrostatic bearing

A variety of designs are used for the capillary resistors, depending upon constructional and performance limitations. Short capillary tubes, which must have a very small bore to provide the necessary resistance, may be accommodated within the supply lines as shown in the upper part of Fig. 4.56. Due to the danger of blockage, the bore size is governed by the largest size of foreign

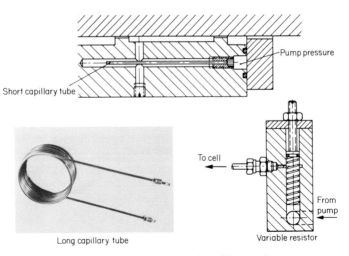

Fig. 4.56 Various designs of capillary resistors

or dirt particles which may be expected in the oil. Larger-bore capillary tubes are formed into spiral coils. In the design of the variable resistor shown on the right of the diagram, a set-screw is used to adjust the resistance. The resistance is set with the aid of a setting jig.

Apart from the use of resistors with a constant resistance value (constant resistors), others are used where the resistance is varied according to the applied load. This permits a marked improvement of the load-carrying capacity and, in particular, the stiffness of a bearing cell, because the resistance reduces as the load (i.e. p_C) increases. Figure 4.57 shows two designs for such load-dependent resistors. This type of resistor has been designed so that the oil gap h remains nearly constant under differing loads. From the

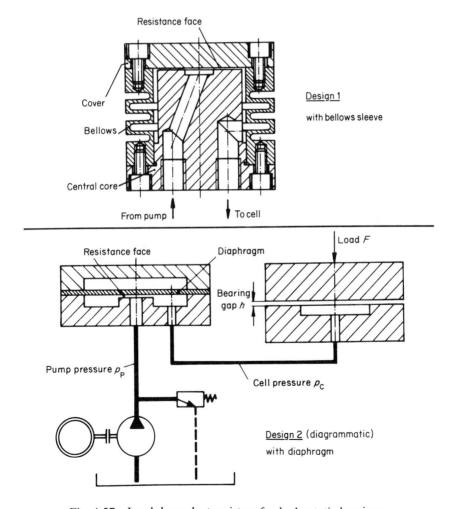

Fig. 4.57 Load dependent resistors for hydrostatic bearings

'Hagen–Poiseuille law' for a bearing cell

$$h^3 = \frac{Q}{p_C} \frac{12\eta l}{b} \tag{4.50}$$

it may be seen that if an oil gap h is to be kept constant, then the flow quantity Q must vary in direct proportion to any change in the cell pressure p_C. In both versions shown in Fig. 4.57 the operating principle is as follows. When the load F increases, the cell pressure p_C rises. This causes the bellows or the diaphragm to expand. The flow resistance reduces, permitting a greater quantity of oil to flow—notwithstanding the heavier load on the bearing. Conversely, the increased resistance of unloaded cells reduces the total oil flow quantity.

Figure 4.58 compares the gap size for varying loads for three systems of oil supply, viz: 'one pump per cell', 'single pump with capillary resistor' and 'single pump with diaphragm resistor'. It may be seen that when the last-mentioned type, where the diaphragm is a load-reactive resistor, is correctly applied, near-ideal conditions are produced, i.e. there is no displacement due to load variations within a given range.[47] However, the application of load-responsive resistors is only of value if the bearing components are correspondingly stiff.

A reduced stiffness but a better load-carrying capacity is found in the 'one pump per cell' system. It is used when heavy overloads on the bearing may be expected. The 'single pump with capillary resistor' system has the least stiffness, but is widely used in practice for economic reasons. If it is carefully planned, a reasonable bearing stiffness may be achieved—even with this system.

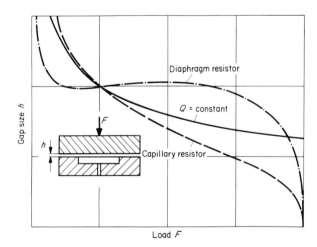

Fig. 4.58 Lubrication gap variation for different oil supply systems without retaining plate

4.2.1.2 Bearing calculations

The designer is given a considerable degree of freedom when employing hydrostatic bearings. His aim will be for the bearing to cope with the expected loads and to provide the necessary stiffness with the minimum pumping power and frictional losses.

The following design and checking formulae may be derived from the definitions made thus far.[49]

Oil supply system 'one pump per cell' (Q_P = constant) without a retaining plate. Considering one cell:

$$p_P = p_C = \frac{F_o + F}{A_{\text{eff}}} \tag{4.51}$$

where F_o is the load at the outlet (e.g. the load due to the weight of slide) and F is the additional machining force.

From equation (4.46) we have

$$Q_P = \frac{p_C}{R_C} = \frac{F_o - F}{A_{\text{eff}}} \frac{bh^3}{12\eta l} \tag{4.52}$$

By transposing the above we obtain the load–displacement relationship:

$$F + F_o = \frac{12\eta l A_{\text{eff}} Q_P}{b} \frac{1}{(h_o - x)^3} \tag{4.53}$$

In the above, $(h_o - x)$ has replaced h, where h_o is the outflow gap height which is formed as a result of an outflow load F_o and x is the displacement of the guide elements under a load F.

If equation (4.53) is applied to the outflow condition which is formed due to the load F_o, i.e.:

$$F_o = \frac{12\eta l A_{\text{eff}} Q_P}{bh_o^3} \tag{4.54}$$

then we obtain the equilibrium condition:

$$\frac{F + F_o}{F_o} = \frac{1}{(1 - x/h_o)^3} \tag{4.55}$$

or

$$\frac{F}{F_o} = \frac{1}{(1 - x/h_o)^3} - 1 \tag{4.56}$$

where:

$$h_o = \sqrt[3]{\left(\frac{12\eta l A_{\text{eff}} Q_P}{bF_o}\right)} \tag{4.57}$$

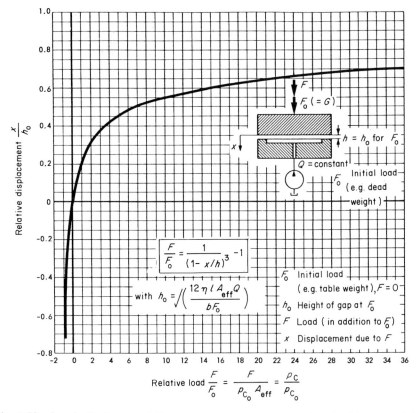

Fig. 4.59 Load–displacement diagram; system: one pump per cell without retaining plate

Figure 4.59 shows the relationship between the load and displacement based on equation (4.56).

The non-linear stiffness relationship can be obtained from equation (4.56):

$$k = \frac{dF}{dx} = \frac{d\{F/F_o\}}{d\{x/h_o\}} \frac{F_o}{h_o}$$

or

$$k = \frac{3F_o/h_o}{(1 - x/h_o)^4} \qquad (4.58)$$

At the origin (i.e. $F = 0$, $x = 0$) the stiffness is

$$k_{x=0} = 3\frac{F_o}{h_o} \qquad (4.59)$$

It may be noted from equations (4.58) and (4.59) that to obtain an adequate stiffness a correspondingly high outflow load F_o will be necessary.

The outflow gap h_o should be kept as small as possible. Its minimum size is, however, governed by the accuracies obtainable during manufacture and the elastic deformations of the machine components, as a metal-to-metal contact between the components must be avoided at all costs. The bearing gap should reduce to 0.3 to 0.5 h_o under maximum load F_{max}. As a guide the following may be used as an approximation:

$$10 \ \mu m < h_o < 60 \ \mu m$$
$$\text{bearing} \qquad \text{guideways}$$

With the aid of equations (4.56), (4.57) and (4.59), as well as the relationships depicted in Fig. 4.59, the design of the bearing can be finalized.

The output power required from the pump will be:

$$P_{P\,max} = \frac{p_{P\,max} Q_P}{\varepsilon_P} \qquad (4.60)$$

Oil supply system 'single pump with capillary resistors' p_P = constant without a retaining plate. From Fig. 4.51 the following is valid for one cell:

$$\frac{p_C}{p_P} = \frac{R_C}{R_C + R_K} \qquad (4.61)$$

$$\frac{F_o + F}{A_{eff} p_P} = \frac{1}{1 + R_K/R_C} \qquad (4.62)$$

If equation (4.62) is applied to the outflow conditions which occur under the outflow load (F_o, h_o) then we have:

$$\frac{F_o + F}{F_o} = \frac{1 + R_K/R_{C_o}}{1 + R_K/R_C} \qquad (4.63)$$

where $R_{C_o} = \dfrac{12 \eta l}{b h_o^3}$

The ratio of the resistances $\xi = R_K/R_{C_o}$ is known as the resistor characteristic. After some transposing, we obtain the following relationship between load and displacement:

$$\frac{F}{F_o} = \frac{1 + \xi}{1 + \xi(1 - x/h_o)^3} - 1 \qquad (4.64)$$

and from equation (4.61):

$$p_{C_o} = p_P \frac{1}{1 + \xi} \qquad (4.65)$$

or

$$\frac{F}{A_{eff} p_P} = \frac{1}{1 + \xi(1 - x/h_o)^3} - \frac{1}{1 + \xi} \qquad (4.66)$$

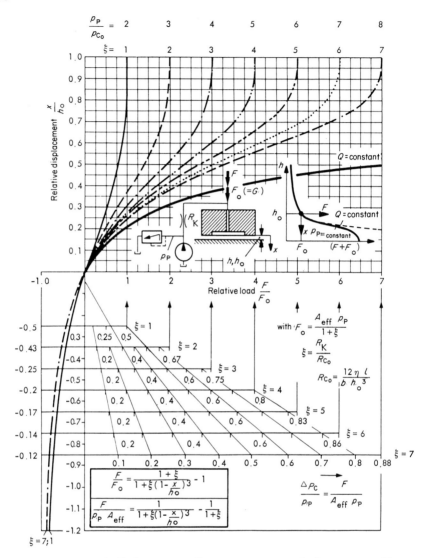

Fig. 4.60 Load–displacement diagram; system: single pump with capillary resistors and without retaining plate.

Figure 4.60 depicts diagrammatically the relationships expressed in equations (4.64) and (4.66). It may be seen from the diagram that the high degree of stiffness in the system 'Q = constant' can only be approached by the system 'p = constant' with high resistances. However, these will require high pump powers (see equation 4.69).

The stiffness is obtained from equation (4.64) as:

$$k = \frac{dF}{dx} = 3 \frac{F_o}{h_o} \frac{(1 + \xi)\xi(1 - x/h_o)^2}{[1 + \xi(1 - x/h_o)^3]^2} \qquad (4.67)$$

At the origin ($F = 0$, $x = 0$) the stiffness is:

$$k_{x=0} = 3 \frac{F_o}{h_o} \frac{\xi}{1 + \xi} \tag{4.68}$$

The pump power is obtained from equation (4.65) as:

$$P_P = \frac{Q_P p_P}{\varepsilon_P} = \frac{p_P^2}{(R_K + R_{C_o})\varepsilon_P} = \frac{p_{C_o}^2(1 + \xi)^2}{(R_K + R_{C_o})\varepsilon_P}$$

$$= \left(\frac{F_o}{A_{\text{eff}}}\right)^2 \frac{(1 + \xi)^2}{R_K + R_{C_o}} \frac{1}{\varepsilon_P} \tag{4.69}$$

Other pump data are:

$$Q_P \geq 1.2 \frac{p_P}{R_K + R_{C_o}} \tag{4.70}$$

$$p_{P\,\text{max}} \geq 1.5 p_{C\,\text{max}}(1 + \xi) \tag{4.71}$$

Oil supply system 'single pump with capillary resistors' p_P = constant use of a retaining plate enables a high pre-load F_o to be applied when desirable. For this reason, as well as the 'clamping effect', the stiffness is considerably improved compared with that of the oil supply system without a retaining plate. Moreover, the guides may carry loads in both directions.

Using the equations introduced above, we obtain the following relationship for the load-displacement conditions:

$$\frac{F}{F_o} = \frac{1}{(1 - x/h_{o_1})^3} - \frac{1}{(1 + x/\lambda h_{o_1})^3} \tag{4.72}$$

where λ is the gap ratio shown on Fig. 4.62. Let the suffix 1 refer to cell 1 and the suffix 2 refer to cell 2. Then the following values apply. The pre-load is:

$$F_o = \frac{12 \eta l_1 Q_1 A_{\text{eff}_1}}{b_1 h_{o_1}^3} = \frac{12 \eta l_2 Q_2 A_{\text{eff}_2}}{b_2 h_{o_2}^3} \tag{4.73}$$

The outflow gap ratio is:

$$\lambda = \frac{h_{o_2}}{h_{o_1}} = \sqrt[3]{\left(\frac{l_2}{l_1} \frac{Q_2}{Q_1} \frac{b_1}{b_2} \phi\right)} \tag{4.74}$$

where ϕ is the area ratio:

$$\phi = A_{\text{eff}_2}/A_{\text{eff}_1}. \tag{4.75}$$

The equation (4.72) is presented diagrammatically in Fig. 4.61. The stiffness may be expressed by:

$$k = \frac{dF}{dx} = 3 \frac{F_o}{h_{o_1}} \left[\frac{1}{(1 - x/h_{o_1})^4} + \frac{1/\lambda}{(1 + x/\lambda h_{o_1})^4}\right] \tag{4.76}$$

At the start ($F = 0, x = 0$) the stiffness is:

$$k_{x=0} = 3\frac{F_o}{h_o}\left(1 + \frac{1}{\lambda}\right) \qquad (4.77)$$

Figure 4.62 shows the stiffness curve as a function of the relative displacement x/h_{o_1} and the gap ratio λ.

Oil supply system 'single pump with capillary resistors' (p_P = constant) with retaining plate. By applying the relationship established for the system with this single pump with capillary resistors above, we obtain the following load-displacement equation for a pair of hydrostatic retaining plate cells:

$$\frac{F}{p_P A_{\text{eff}_1}} = \frac{1}{1 + \xi_1(1 - x/h_{o_1})^3} - \frac{\phi}{1 + \xi_2(1 + x/\lambda h_{o_1})^3} \qquad (4.78)$$

and when related to the pre-load F_o:

$$\frac{F}{F_o} = \frac{1 + \xi_1}{1 + \xi_1(1 - x/h_{o_1})^3} - \frac{1 + \xi_2}{1 + \xi_2(1 + x/\lambda h_{o_1})^3} \qquad (4.79)$$

Let the index 1 refer to cell 1 and the index 2 refer to cell 2. Then the following values apply. The pre-load is:

$$F_o = \frac{A_{\text{eff}_1} p_P}{1 + \xi_1} = \frac{A_{\text{eff}_2} p_P}{1 + \xi_2} \qquad (4.80)$$

The outflow gap ratio is:

$$\lambda = \frac{h_{o_2}}{h_{o_1}} \qquad (4.81)$$

The area ratio is:

$$\phi = \frac{A_{\text{eff}_2}}{A_{\text{eff}_1}} \qquad (4.82)$$

and the resistor ratios are:

$$\xi_1 = \frac{R_{K_1}}{R_{C_{o1}}} \quad \text{and} \quad \xi_2 = \frac{R_{K_2}}{R_{C_{o2}}} \qquad (4.83)$$

The following functional relationship exists between both resistor ratios and the effective cell area ratio:

$$\phi = \frac{\xi_2 + 1}{\xi_1 + 1} \qquad (4.84)$$

The equation (4.78) is presented graphically in Fig. 4.63. Tests have shown that the stiffness characteristics are at an optimum when $\xi_2 \approx \phi$; therefore this condition is assumed in Fig. 4.63.

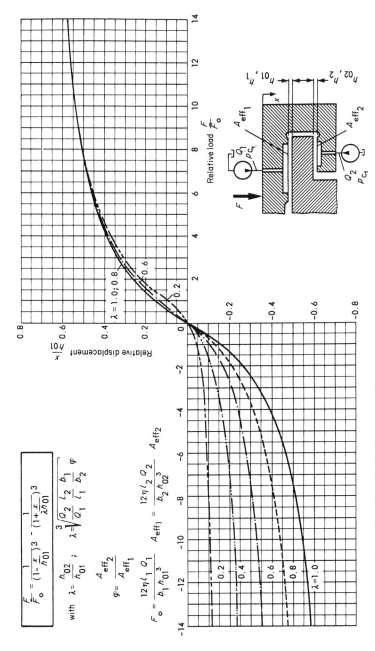

Fig. 4.61 Load–displacement diagram; system: one pump per cell with retaining plate

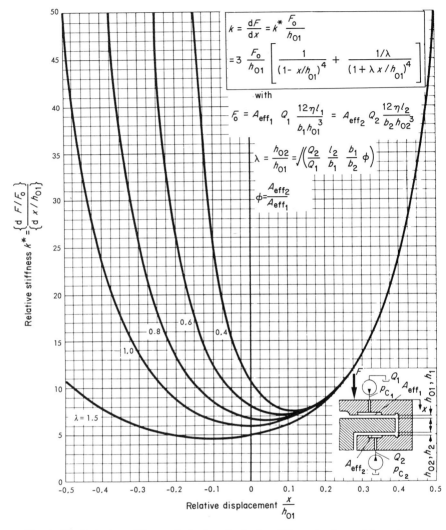

Fig. 4.62 Relative stiffness plotted against x/h_{o1} different values of λ; system: one pump per cell with retaining plate

It may be noted when comparing Figs. 4.61 and 4.63 that the stiffness characteristics of the two systems differ mainly at the higher loads. Whereas the stiffness in the system where $p = $ constant approaches zero at the higher loads, in the case of the system where $Q = $ constant the stiffness tends to infinity as the load is increased.

The stiffness relationship is obtained from a complicated formula; therefore in this case only the stiffness at the origin ($F = 0, x = 0$) is quoted here:

$$k_{x=0} = \frac{dF}{dx} = 3 \frac{F_o}{h_{o1}} \left(\frac{\xi_1}{1 + \xi_1} + \frac{\xi_2/\lambda}{1 + \xi_2} \right) \quad (4.85)$$

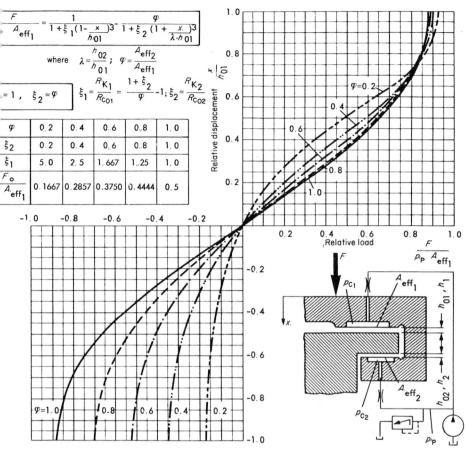

Fig. 4.63 Load–displacement diagram; system: single pump with capillary resistors with retaining plate ($\lambda = 1$; $\xi_2 = \phi$)

Figure 4.64 shows this stiffness characteristic based on equation (4.85) as a function of the outflow gap ratio λ and the cell area ratio ϕ.

4.2.1.3 Dynamic behaviour

The dynamic behaviour of a system is in general described by considering stiffness, damping and mass. With regard to stiffness, a differentiation must be made between the static stiffness, which is governed by the degree of control of the cell pressure over the height of the oil film gap, and the stiffness of the oil column in the supply lines and cells. The stiffness of the oil column is of no significance for the static behaviour, because it is stabilized over the oil film gap for slow movements. Only at higher frequencies is such a consideration necessary, when there is inadequate time between two movements for a levelling out of the cell pressure.

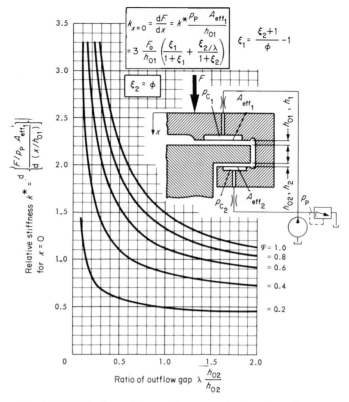

Fig. 4.64 Relative stiffness when $x = 0$ plotted against λ for different values of ϕ; system: single pump with capillary resistors with retaining plate

Damping results from the relative motions over the bearing lands (squeeze film), as well as in the oil supply lines and resistances. Movements at right angles to the cell provide a much greater damping effect than that obtained with rolling bearings. Motions parallel to the cell give little damping as there is no oil displacement in that case.

For stiff spindle bearing units, as shown in Fig. 4.65, the vibration nodes of the spindle occur close to the bearing positions. In such cases, there is only minimal relative motion, and hence little effective damping force. This situation may be improved by taking special steps to reduce the stiffness of the oil column. Such efforts have no influence on the static behaviour as the static displacements are controlled by the oil film gap. However, with regard to the dynamic behaviour, increased damping forces reduce the vibration amplitudes.

Figure 4.65 shows a practical application of these considerations.[47,50] In the front radial bearing, each cell has a hole drilled into it which is sealed with a screw cap. The cap has a number of steel diaphragms soldered to it, which

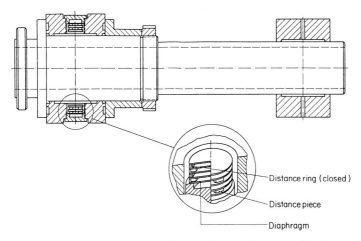

Fig. 4.65 Hydrostatic spindle bearing unit with diaphragm absorption units in the cells of the front radial bearing

move in response to any increase in pressure, and thus effect a reduction in the stiffness of the oil column.

To increase their storage capacity, the diaphragms are so arranged that each pair connected with a distance ring forms a further storage unit. The individual storage units are connected to each other by distance pieces so that the oil can enter into the space between them without hindrance. The expansion of the complete unit corresponds to the increase in the compressibility of

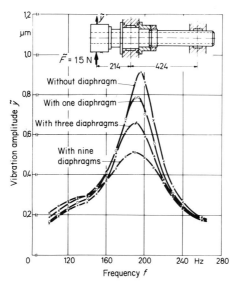

Fig. 4.66 Resonance curves for a spindle bearing using different numbers of diaphragm absorption units

the oil. The effect of such an arrangement is illustrated by the curves in Fig. 4.66. The vibration amplitudes are shown in the range of the dominant resonance frequency and depend upon the number of diaphragm absorption units. A marked reduction of the resonance amplitude near the chuck is achieved with increasing numbers of diaphragms, i.e. an increase in the absorption volume.

4.2.1.4 Energy consumption and hydraulic circuits

Energy consumption. The energy consumption of a hydrostatic bearing is dependent upon the frictional work done and the work done by the pump, as shown in Fig. 4.67. The addition of these losses is the total power loss in the bearing, which is entirely converted into heat.[47] As previously described for the behaviour of plain bearings, the hydrodynamic friction force is, according to Newton's law, given by:

$$F_R = A_R \eta \frac{v}{h} \tag{4.86}$$

where A_R = area of the bearing cell land
 η = oil viscosity
 v = sliding velocity
 h = height of the oil film gap

In contrast to the frictional areas of hydrodynamic plain bearings, in hydrostatic bearings the areas of the cells are of no importance as the cell depths are much greater than the height of the oil gap, hence, in this respect, their frictional force may be ignored. The frictional work done is given by:

$$P_R = F_R v$$
$$= A_R \eta \frac{v^2}{h} \tag{4.87}$$

The second constituent of the energy loss—the work done by the pump—is obtained from the product of the flow quantity and the pump pressure,

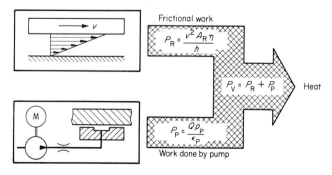

Fig. 4.67 Origins of losses in a hydrostatic bearing

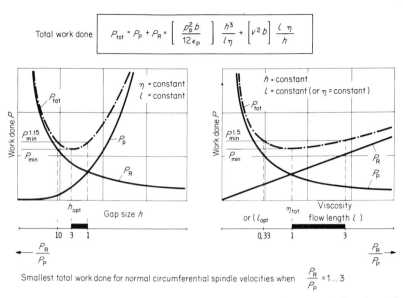

Fig. 4.68 Optimization of the total work done for a hydrostatic bearing oil supply system where Q = constant

divided by the pump efficiency, i.e.:

$$P_P = \frac{Q p_P}{\varepsilon_P} \qquad (4.88)$$

In order to minimize the energy consumption and hence the heat generated by the bearing, the total power loss must be kept as low as possible. The total power loss is depicted in Fig. 4.68 as a function of the height of the lubricating film gap h on the left, and on the right as a function of the viscosity η. The equation used for P_{tot} in these curves is given in the upper part of the diagram. The value of the flow quantity Q for the work done by the pump is obtained by application of the Hagen–Poiseuille law; the frictional area A_R for the work done by friction is the area of the cell land, determined from the cell geometry. If the characteristics of the oil supply system and the sliding velocity v are kept constant, then the total work done (considered from the viewpoints of the gap size h and the oil viscosity η) may be minimized. On the left diagram, the optimum gap size h_{opt} is indicated where the value P_{tot}, being the sum of P_P and P_R, is a minimum. Similarly, on the diagram on the right, the optimum viscosity η_{opt} or the optimum flow length l_{opt} is indicated for a minimum P_{tot}.

For machine-tool spindles running at normal circumferential velocities v, the minimum total work done is usually obtained when the ratio of frictional work done to pump work is between 1 and 3.

Hydraulic circuit. The supply of pressurized oil to the cells and the return of the oil flowing out of the cells constitute the hydraulic circuit. It should satisfy

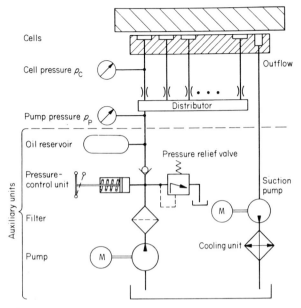

Fig. 4.69 Hydraulic circuit and accessories of a hydrostatic bearing

the following criteria:

(a) an adequate supply of pressured oil to the bearing;
(b) the removal of heat generated due to power losses;
(c) the prevention of physical contact in the bearing in the event of pump failure.

A hydraulic circuit diagram for the oil supply system 'single pump and capillary resistors' is shown in Fig. 4.69. The individual cells are fed from a distributor through capillary tubes. A constant pressure p_P is provided by means of an oil reservoir, pressure control unit, pressure relief value, filter and pump. All kinds of pumps are used including axial and radial piston pumps, butterfly-cell pumps, gear pumps and screw pumps. The most popular is the low-cost gear pump. A cooling unit is fitted behind the suction pump for any heat dissipation which may be necessary. To prevent possible seizure in the event of pump failure, the oil contained in the reservoir provides an emergency supply to the bearing.[47] The reservoir also smoothes the oil flow, e.g. in the event of any pulsation effect from gear pumps, and thus ensures a stable bearing operation.

4.2.2 Hydrostatic linear bearings

The principles of design and construction of hydrostatic slideways are not fundamentally different from conventional types. Consequently, only the special design features, examples of application and means of compensation for guide inaccuracies are discussed here.

4.2.2.1 Design features and general arrangements
A number of cells are provided for guideways which are situated parallel to the direction of movement. The arrangement of cells for a machine-table guide is shown in Fig. 4.70. At least two cells must be provided on each guideway to absorb torque. A larger number of cells evens out any waviness in the guideways and is better able to compensate for any deviation from straightness in the table. Hence, large numbers of smaller cells are recommended.

There are various geometric forms of cells in use. Three designs are shown in Fig. 4.71.[47] Experience has shown that the cell depth should be between 10 and 100 times greater than the oil film gap, i.e. 0.5 to 5 mm, which will provide an even pressure distribution within the cell and keep the frictional work to a minimum. For low sliding velocities, a groove as shown on the right in the diagram is frequently adequate. The advantage of this design lies in the feature that in the event of pump-pressure failure better emergency conditions exist and thus the damage will be minimized in any eventual metal-to-metal contact. Sharp-edged cells should be avoided owing to the dangers from dirt particles. Furthermore, experience has shown that the outflow length l on guides should be between a third and a fifth of the total width of a cell.

Apart from the differing cell shapes, a range of cell types may be defined according to the outflow conditions. The cell types depend on the order in

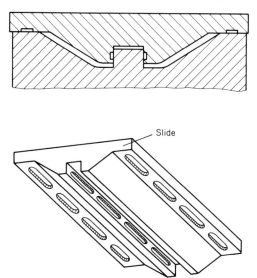

Fig. 4.70 Arrangement of cells for hydrostatic machine table guideway

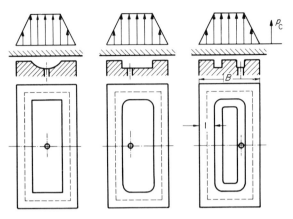

Fig. 4.71 Various shapes of cells for hydrostatic bearing surfaces

which the cells are arranged upon the guideways, as clarified in Fig. 4.72. The free-standing cell shown as type I on the left of the diagram allows the oil to flow out over all the lands. This is made possible by the return-flow grooves between the cells, and a larger volume of oil flow will be used for a given load-carrying capacity. This type of cell is only used in special cases. The cell type II is an end cell of a series. The oil outflow occurs over three lands only. The inner cell within a row, marked type III, has its oil outflow only over the two longitudinal lands. The oil flow from one cell into the other may be

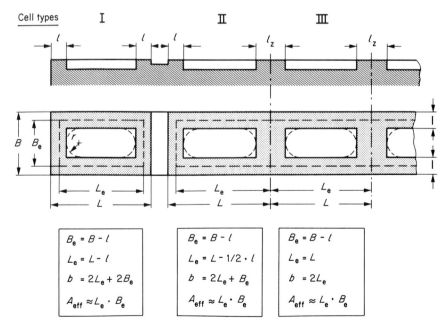

Fig. 4.72 Various cell types for hydrostatic bearing

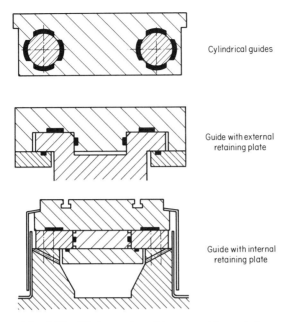

Fig. 4.73 Hydrostatic guideways with retaining plates

ignored in these cases, because the adjacent cells have a similar pressure distribution. In the lower part of the diagram, the values for the outflow widths and the effective flow areas are given for the three cell types.

In general, hydrostatic slides are fitted with retaining plates. Three different examples of designs for slideway guides are presented in Fig. 4.73. In the upper drawing a cylindrical guide is shown. The advantage here is the simple construction, but considerable bending effects may occur on smaller-diameter circular slides. Moreover, thermoelastic deformations cannot be accommodated due to the static restriction in movement. The guideway shown in the centre of the illustration with the external captive plate has wide application because it is easily assembled. However, the third guide shown with its internal retaining plate, which is subjected to a reduced bending effect, has the advantage of providing a comparatively better stiffness.

In section 4.2.1 the displacements and stiffness for a single cell were described. A larger number of cells together with several guideways produce a static system with a number of unknowns for which calculations are very difficult. To simplify such mathematical analysis, it is assumed that the machine bed and table are both rigid and that all deformations may be traced back to changes in the lubrication film gap.

4.2.2.2 Application principles
Figure 4.74 shows a section through a hydrostatic machine-tool slide. To calculate the displacements of the table, appropriate computer programs must

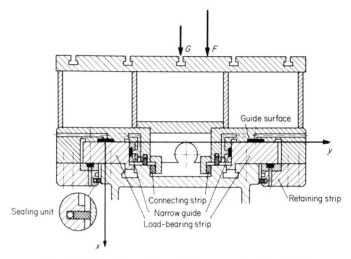

Fig. 4.74 Hydrostatic machine-tool slide (section)

be used. By considering all the cells the equilibrium conditions are established, from which the local deformations and tilting of the slide can be forecast. Frequently, the elastic behaviour of the machine-frame structure and the adaptation of components must be included in the calculations (see sec-

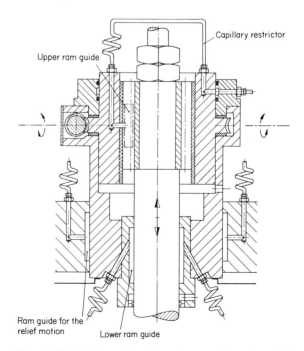

Fig. 4.75 Hydrostatic guides of ram and ram head of a gear-shaping machine (Liebherr)

tion 2.6.2), as only then can the effective heights of the clearance gaps be established and the total displacement determined.

A further example of application is the hydrostatic guide for the ram and ram-head of a gear-shaping machine, shown in Fig. 4.75.[48] The hydrostatic slides are in this case particularly advantageous as they obviate the continuous changes from static to motion friction conditions. The upper guide serves for the linear up and down movement of the ram. This guide, shown in section in Fig. 4.76, also accommodates the rotary motion induced by a worm and worm wheel. The lower guide consists of a simple circular guide for the ram and a flat guide for the ram-head. The flat guide is necessary to support the relief clearance motion of the cutter during the return stroke. (A detailed description of the mode of action is given in Volume 1, with a description of the gear-shaping machines.)

The sectioned drawing in Fig. 4.76 shows a guide which is made as a straight spur gear with every third tooth removed to suit the cell arrangement; the space vacated is shaped to suit the bearing ring. The accuracy of the internal bore of the sleeve is achieved by casting with a plastic material. The

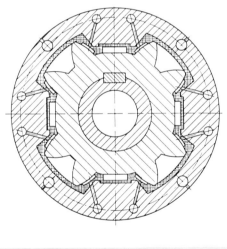

Fig. 4.76 Hydrostatic ram guide (Liebherr)

cells are made by fixing foil to the tooth flanks of the spur gear with an adhesive to act as patterns during casting. After extracting the spur gear from the bearing sleeve, the foil patterns are removed from the solidified plastic. The gap size for the oil film is determined by the shrinkage of the cast plastic.

4.2.2.3 Compensations for guide inaccuracies

As indicated for hydrodynamically lubricated guideways, inaccuracies are normally present in a guiding system. They may be traced to errors in manufacture, to wear or to deformations of the guideway due to external forces or temperature fluctuations. Such inaccuracies may be compensated by control of the cell pressure. The principles underlying such bearing control are outlined in Fig. 4.77. The vertical support of the slide is on four cells, each of which has its own control circuit.[51] All control circuits are connected to one overall system. The reference surface for the pneumatic measurement of the actual table disposition consists of two straight edges. At four points the actual pressure against the straight edge is monitored by a pneumatic transducer and compared against a nominal pressure which corresponds to the correct slide position. The differential between nominal and actual pressures operates a diaphragm valve which regulates the oil flow. This modifies the gap height, thus compensating deviations in the table position. The pneumatic system using a transducer and straight edge may be replaced by an electronic control using laser beams and photoelectric cells; the straight edge is then replaced by a reference laser.

The improvement of a slide's guideway by a compensating bearing control is shown in the curves of Fig. 4.78. The movement of the slide tilts the table about an angle α, which is affected by the load. The high degree of scatter

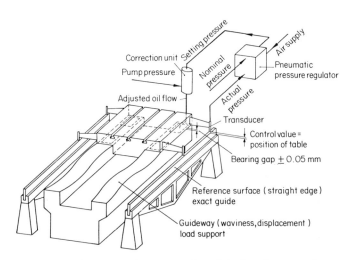

Fig. 4.77 Diagrammatic presentation of the bearing control on a machine tool

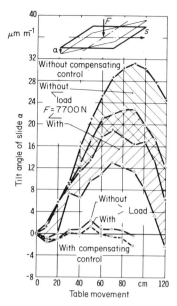

Fig. 4.78 Tilt of slide in relation to distance moved

indicated is caused by operating conditions and may be traced to thermal influences. A marked improvement is achieved by the compensating control. The system described here is capable of compensating for linear bearing errors of up to ±60 μm and up to a frequency of 1 Hz.

4.2.3 Hydrostatic plain circular bearings

Hydrostatic bearings for spindles in machine tools occupy a special position alongside rolling and hydrodynamic bearings. This is due to combination of the functional principles to provide unique characteristics which can be adapted in most instances to the particular requirements of a given situation.

Apart from the fact that hydrostatic bearings are virtually wear-free, they also run with little noise, have a wide speed range, as well as good stiffness combined with high damping effects. However, as with all hydrostatic units, the costs of the oil supply and safety systems are high.

4.2.3.1 Design variations

A variety of basic designs of hydrostatic plain cylindrical bearings is presented in Fig. 4.79. If the classification is based on the main direction of loading, then three groups may be identified:[52]

(a) radial bearings;
(b) combined radial and thrust bearings;
(c) thrust bearings.

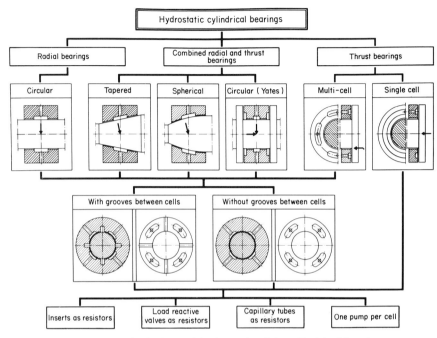

Fig. 4.79 Basic designs of hydrostatic plain cylindrical bearings

Radial bearings are mostly circular or, occasionally, for ease of assembly incorporate slightly tapered bearing surfaces, and are only capable of supporting loads which are applied perpendicular to the spindle axis and at the centre of the bearing. Combination bearings, as shown in the centre of the diagram, may be additionally loaded in the direction of the spindle axis. A further subdivision identifies: (1) tapered bearings, which are mainly used when the provision of a separate thrust bearing is either not possible, e.g. due to lack of space, or is considered unnecessary because the thrust is not high; (2) spherical bearings which perform similarly to tapered bearings but are preferred when a high degree of misalignment is likely between the spindle and bearing axes; and (3) the so-called Yates bearing which is a combination of a circular radial bearing and a plain thrust bearing where the oil flowing out at the sides of the circular bearing is utilized to supply the cells of the thrust bearing. None of the combination bearings can provide good stiffness and load-carrying capacity in both extreme loading directions. The third group are the plain thrust bearings which can only absorb loads parallel to the spindle axis.

A further subdivision is made depending upon whether or not oil return-flow grooves between the cells are provided. For a given load, bearings without grooves provide better stiffness than those where such slots are present, but the latter are capable of a better heat dissipation due to the larger rate of oil flow. Finally, any one of the oil supply systems shown at the bottom of Fig. 4.79 may be used, i.e. a single pump with restrictor inserts, load reactive valves or capillary tubes as resistors, or 'one pump per cell' systems.

4.2.3.2 Pressure build-up

Figure 4.80 shows the pressure gradient in the cells of a circular radial bearing with oil return grooves. The upper diagram illustrates the pressure gradient both in the cross-sections and longitudinally along the cells of an unloaded bearing. From the lower diagram it may be seen that when a load F is applied on the centre-line of a pocket, the effect on the cells in line with the force

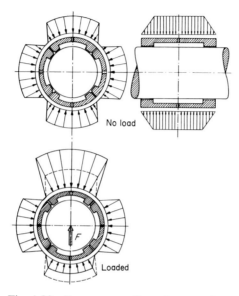

Fig. 4.80 Pressure gradient in a radial bearing with oil-return grooves between cells

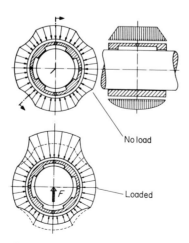

Fig. 4.81 Pressure gradients in a radial bearing without oil-return grooves between cells

produces a pressure gradient similar to that of a guideway fitted with a captive plate, whilst the cell pressures perpendicular to the applied load are unaffected.

The oil return grooves which separate neighbouring cells in the circumferential direction permit the pressurized oil to flow over all the four lands of a cell, axially as well as along the circumference, with the pressure falling to zero on the external side of a land. In contrast, Fig. 4.81 shows the pressure gradient in a bearing where there are no oil return grooves between the cells. In this type, which is mostly applied today, the oil can only flow out in the axial direction, making cell pressures independent of each other.

4.2.3.3 Bearing designs

The design and mathematical analysis of hydrostatic axial bearings does not substantially differ from the principles described in section 4.2.1 for linear guides. For this reason the mathematical analysis of radial bearings is not dealt with further here. Each bearing surface of a radial bearing is provided either with a single annular cell or with a number of cells (see Fig. 4.79).

The conditions in the fluid-film gap of a hydrostatic radial bearing are most complicated. The analogy of an electric circuit diagram, depicted in Fig. 4.82 for a bearing with four cells and loaded in the direction of the lands, may be used for the design of the bearing as well as the oil supply system. In the system shown on the left of the diagram, i.e. single pump with capillary resistors, the oil flows firstly through the resistors R_K. After that, the flow may be either in a circumferential direction through resistances R_u into the neighbouring cells or through resistors R_a in both axial directions. <u>As the pressure</u>

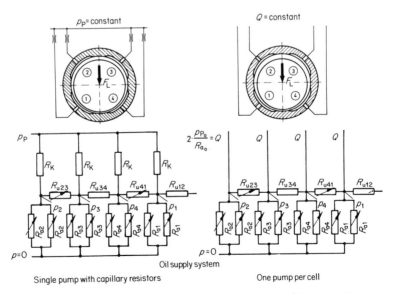

Fig. 4.82 Electric circuit analogy for bearings with four cells

differences are the same, the axial resistances may be considered to be in parallel. In the 'one pump per cell' system, similar conditions occur, without the capillary resistors. With the aid of this analogous circuit and the flow equations for parallel gaps and capillary resistors, the flow quantity and the bearing clearances under load may be calculated.[52]

For a bearing with oil return grooves between the cells, as shown in Fig. 4.80, the hydraulic resistance of each cell is a parallel combination of the resistances R_u in the circumferential direction and R_a in the axial direction, both being at atmospheric pressure when $p = 0$.

When designing a bearing, it is also important to establish its stiffness in addition to the determination of its load-carrying capacity. Figures 4.83 to 4.85 present the load–displacement relationships for a number of radial bearings with different oil supply systems. The curves have been established using the analogies described above for a load applied in the direction of the centre of the land as well as the centre of the cell.

The following nomenclature has been used in Figs 4.83 and 4.84:

p_P = pump pressure
D_L = bearing diameter (shaft)
L_e = effective length of bearing
h_o = bearing clearance when $F = 0$, ignoring the weight of the shaft
ξ = resistance ratio, $\xi = R_K/R_{C_o} = 1$ (in this case)
R_K = capillary resistance, $R_K = 8\eta l_K/(\pi r^4)$ (see equation 4.49)
R_{C_o} = cell resistance for h_0 ($R_{C_o} = R_{a_o}$ for cells without connecting grooves)
R_{a_o} = axial outflow portion of cell resistance,

$$R_{a_o} = \frac{12\eta l_a}{b_a h_o} \qquad (4.89)$$

R_{u_o} = circumferential outflow portion of cell resistance,

$$R_{u_o} = \frac{12\eta l_u}{b_u h_o^3} \qquad (4.90)$$

κ = resistance ratio of axial (R_{x_o}) to circumferential (R_{u_o}) portion of cell resistance for gap height, h_o

$$h_o = \frac{R_{a_o}}{R_{u_o}} = \frac{l_a}{b_a} \frac{b_u}{l_u} \qquad (4.91)$$

x = displacement of shaft due to load F

Both diagrams are based on an optimum resistance ratio, $\xi = 1$. It can be seen from the curves shown that the stiffness increases as the resistance ratio κ decreases.

An advantage of a six-cell bearing over a four-cell bearing is the directional sensitivity of the stiffness, in particular for high κ values. Figure 4.85 illustrates the characteristic curves for a four-cell bearing with an oil supply system where Q = constant, i.e. one pump per cell.

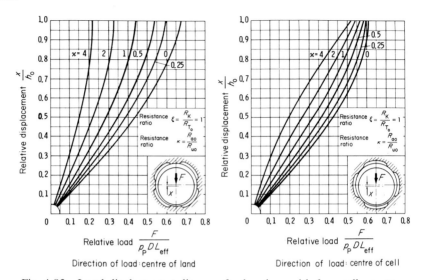

Fig. 4.83 Load-displacement diagram for bearings with four cells; system: single pump with capillary resistors

For an overall calculation, an empirical relationship has been established[47] which is based on these curves and where the lower sections have been approximated to straight lines:

$$\chi = \frac{x}{h_o} = \frac{1}{0.24}\sqrt[3]{\left(\frac{\kappa}{z^2}\right)\frac{F}{p_P D_L L_e}} \qquad (4.92)$$

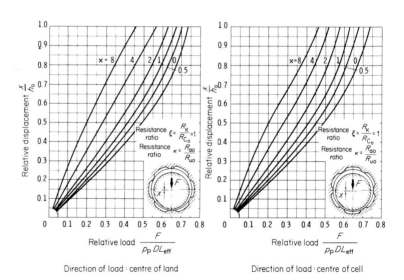

Fig. 4.84 Load–displacement diagram for bearings with six cells; system: single pump with capillary resistors

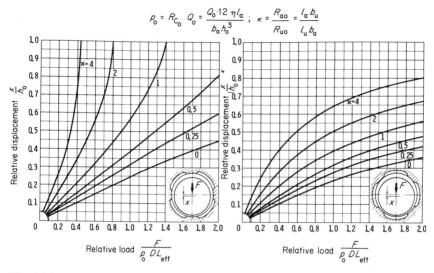

Fig. 4.85 Load–displacement diagram for bearings with four cells; system: one pump per cell

$$k = \frac{dF}{dx} = 0.24 \frac{p_P D_L L_e}{h_o} \sqrt[3]{\left(\frac{z^2}{\kappa}\right)} \qquad (4.93)$$

where z represents the number of cells. The following limitations are applied when this equation is used:

$$\xi = 1, \qquad z \geqslant 4, \qquad \chi \leqslant 0.6, \qquad \frac{z^2}{100} < \kappa < \frac{z^2}{5}$$

4.2.3.4 Seals and sealing

The oil flowing out of the cells, having lost almost all its pressure, must be returned to the oil supply system without undue contamination. For this reason, the outflow channels must be provided with seals. The problems associated with sealing and oil return flow cannot be solved by rigidly applied rules and generalizations; each case must be considered on its own merits.

Figure 4.86 illustrates some methods for sealing hydrostatic bearings. Contacting seals, e.g. the lip seal shown in the upper left of Fig. 4.86, should be avoided, especially at high running speeds, because of the associated friction and wear. With the use of helically grooved shafts and labyrinth glands, contact-free seals can be provided. To improve the sealing effect, auxiliary compressed air may be used, as shown in the lower left of Fig. 4.86. Quite often the oil is drawn out by suction, so that the function of the seal is only to prevent ingress of contaminants into the oil circuit. A jet pump working on the venturi principle may be employed as a suction pump, as shown in the lower right of Fig. 4.86.

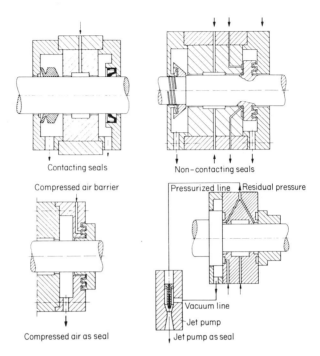

Fig. 4.86 Types of seals for hydrostatic spindle bearings

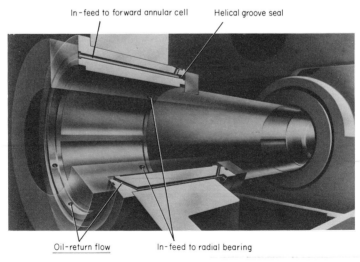

Fig. 4.87 Hydrostatic bearing of a drilling and milling spindle (Wotan)

4.2.4 Hydrostatic spindle bearing systems in machine tools

An important advantage offered by the employment of hydrostatics in bearings of machine-tool spindles is the possibility of optimum bearing characteristics for the most varied service conditions. For low running speeds, hydrostatic bearings are superior to plain hydrodynamic bearings due to the improved load-carrying capacity and friction conditions, whilst at high running speeds and dynamic loading they give better performances when compared with rolling bearings.

In particular, in heavy and in special-purpose machine tools, the main spindles are frequently fitted with hydrostatic bearings. Apart from the design of the bearings, their arrangement and positioning plays an important role. In the case of work-supporting spindles on machine tools, the bearing nearest the work is usually arranged as a fixed bearing. Figure 4.87 shows a hydrostatic bearing of a drilling and milling spindle, with a combined radial and thrust front bearing on the left and the rear radial bearing on the right.

4.2.5 Hydrostatic lead-screws and nuts

Lead-screws and nuts are used to convert rotary motion into linear motion. For low-duty secondary functions hydrodynamic plain lead-screws are used, while for better performance on feed drives recirculating ball lead-screws and nuts are applied. However, especially for large and heavy machine tools and broaching machines, hydrostatic lead-screw nuts are fitted. In these cases they may be used, for example, to drive the machine table. The main advantages of a hydrostatic lead-screw nut when compared with a recirculating ball nut are the better damping properties, the inherent freedom from backlash due to their functional principle and the freedom from wear.

The sectional drawing in Fig. 4.88 shows the principles of a hydrostatic lead-screw nut; the cells are situated in its thread flanks. They are supplied with pressurized oil through capillary resistors, which are also accommodated in the nut. The return flow of the oil is through small holes drilled into the crests and roots of the thread flanks. Due to the large number of cells, e.g.

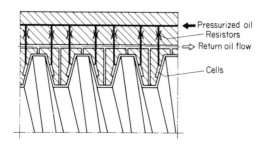

Fig. 4.88 Principle of a hydrostatic lead-screw nut

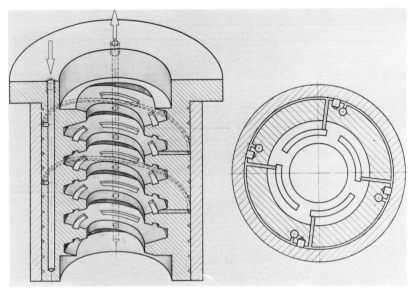

Fig. 4.89 Oil supply for hydrostatic lead-screw nut

four cells for each flank side and thread pitch over a total of between four and eight pitch lengths, the oil supply system 'single pump with capillary resistors' must be used.

Figure 4.89 is used to explain one particular operating principle and the arrangement is of its restrictor resistances.[53,54] On the left of the picture a section through the nut is shown. The oil flows from an axial supply hole through short connecting holes into grooves on the outer circumference. These grooves are cut into the outer sleeve in the form of a helix above the centres of both thread flanks and at the same pitch as the thread. Parts of these grooves provide the flow resistance in conjunction with the close fit of the covering sleeve, and are now referred to as 'groove capillary resistors'. From these, the oil is fed through small holes into the cells. The spiral groove on the outside diameter of the nut is divided into sections with blocking pieces between a connecting hole and the nearest supply hole, so that each cell has its own independent groove capillary resistor. The contact faces are made of a plastic material and are inserted into this large nut during casting with the lead-screw in position. For the formation of the cells, foils are fixed to the thread flanks with an adhesive (see also section 4.2.2.2).

A similar principle is applied in the so-called hydrostatic 'Johnson drive', which is used for the feed mechanism of large machine tools (Fig. 4.90). In this case, a short worm is used, which is not fully enclosed by a nut. Instead, a long rack performs the function of the nut on one side of the worm only. As the worm is short and the rack is firmly fixed to the slide, the axial flexibility of the mechanical components is low. In order to minimize friction between the rack and the worm, the hydrostatic principle is applied. The oil supply to

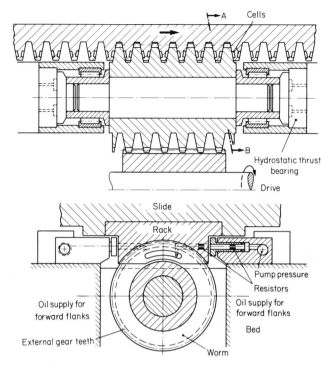

Fig. 4.90 Feed mechanism with hydrostatic rack and pinion system—Johnson drive (Ingersoll)

the cells in the rack is limited to the length covered by the worm, so that oil consumption remains low. The worm is supported by a hemispherical hydrostatic thrust bearing in its axial direction. The worm drive is obtained from a pinion which is in mesh with the teeth on the circumference of the worm.

4.3 Aerodynamic and aerostatic slideways and bearings

In the case of aerodynamic and aerostatic slideways and bearings, air is the separation medium between the sliding surfaces. From this a number of specific characteristics and advantages are obtained. For example, there is no additional constructional expense involved for the provision of a return flow or for seals.

With regard to the mode of operation, there is no fundamental difference from the hydrodynamic and hydrostatic principles; a differentiation is made between aerodynamic and aerostatic bearings based on an analogy with the fluid-lubricated bearings. Both types are frequently simply designated as 'air bearings'. Air bearings exhibit exceptionally low frictional losses, even at very high relative velocities, due to the extremely low viscosity of air; hence they are particularly suitable for fast-running bearings with low loads.[55] Linear air bearings always operate according to the aerostatic principle due to the low sliding velocities. However, there are so far only very few known applications of aerostatic guideways.

4.3.1 Fundamentals and functional principles

The principles underlying air bearings are depicted in Fig. 4.91 with the aid of an aerodynamic and an aerostatic radial bearing. The lubrication medium in an aerodynamic—as in a hydrodynamic—bearing is brought into the reducing clearance gap between the moving parts by a drag effect, resulting in the load-carrying pressure.

Whilst the negative pressure build-up in the widening section of the bearing clearance has no significance in relation to the positive pressure distribution for hydrodynamic bearings, in the case of aerodynamic bearings this vacuum cannot be ignored. In such bearings, the pressures generated in the lubrication gap are in general less than 1 bar, and hence in the same order of magnitude as the negative pressure, whilst in a hydrodynamic lubrication gap, pressures up to 100 bar may be experienced.

An aerostatic bearing and its pressure distribution in service is illustrated on the right of Fig. 4.91. As in the hydrostatic bearing, the shaft in an aerostatic bearing floats on a pressurized medium, the pressure of which is maintained by an external source. Whereas in a hydrostatic radial bearing usually four larger cells are provided in the bearing shell, in an aerostatic bearing one or several nozzle rings are used.

The calculations for the air gap height, air consumption and load-carrying capacity of an aerostatic bearing are more complicated than for a hydrostatic bearing, because a range of additional variables, such as the compressibility of air, flow velocity, turbulence and shape of nozzle, must be allowed for. For the equations for the flow rate in the nozzles and the parallel gap, experiment-

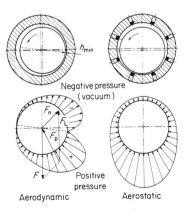

Fig. 4.91 Principles and pressure distribution of aerodynamic and aerostatic bearings (R. Lehmann)

al research results are available for consideration,[55] as well as the laws for boundary layer theory.

4.3.2 Characteristics

The most important advantages and disadvantages of air bearings are set out in Fig. 4.92. Considerable differences are notable when these are compared with oil-lubricated bearings due to the low viscosity and high compressibility of air. The viscosity of air at room temperature is more than three powers of ten lower than that of typical lubricating oils. Consequently, air bearings exhibit very low friction effects, heat generation and damping characteristics. However, the very low viscosity sharply limits the load-carrying capacity because the pressure build-up cannot be very high. An advantage is the great consistency of the viscosity within a wide temperature range, enabling air bearings to be used when large temperature differentials may be expected. Owing to the compressibility of air, air bearings are particularly prone to vibrations.

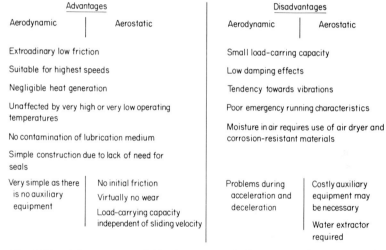

Fig. 4.92 Advantages and disadvantages of air bearings (R. Lehmann)

4.3.2.1 Air consumption and load-carrying capacity

The running costs of an aerostatic bearing is largely governed by the volume of air consumed, which is required to provide a specific load-carrying capacity and/or stiffness. For this reason efforts are made to improve these characteristics by carefully designed bearing clearance gaps. In Fig. 4.93 the pressure distributions over the bearing surfaces and the gap size as a function of the load are shown for three circular flat bearings with differing gap geometry. In the upper part of the diagram the plate bearings (which, for example, might be used as hemispherical bearings) are diagrammatically presented with parallel, concave and variable bearing gaps. Below each sketch the respective

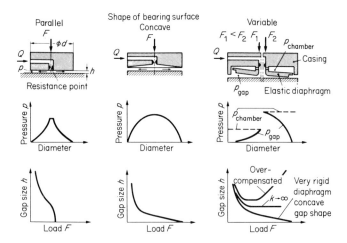

Fig. 4.93 Displacements for aerostatic flat bearing surfaces with varying gap geometrics (Blondeel)

pressure distributions over the bearing surfaces are shown. In the case of the bearing with the parallel gap, the pressure drops rapidly from the point of air input towards the edges of the surface. This tendency is much reduced when the gap geometry is concave, producing a better load-carrying capacity as shown in the lower part of the diagram.

The design incorporating a variable shape of the bearing gap shown on the right of Fig. 4.93 offers the special feature of the ability to compensate for the displacement under load, so that within certain limits a near-infinite stiffness ($k \to \infty$) may be obtained. Such a bearing functions in the following manner. The upper surface of the diaphragm, which has a larger area than that of the bearing, is subjected to the pressure acting at the centre of the bearing, i.e. $p_{chamber} = p_{Gap\ max}$. This causes the pre-loaded diaphragm to change from its original convex shape, so that a concave gap geometry is assumed. If the supply pressure, stiffness and the effective surfaces of both sides of the diaphragm are carefully balanced, then it is possible that the bearing housing will not be deformed within a wide range of increasing load. In such cases, the term 'complete compensation' is applied, which is designated $k \to \infty$ on the lower right of Fig. 4.93. Incorrect balancing of the parameters may lead to overcompensation and consequent instability, as shown on the upper curve.[56]

4.3.2.2 Dynamic behaviour

The dynamic behaviour of air bearings presents problems due to the low damping characteristics and the compressibility of air. A number of significant effects occur which were negligible in hydrodynamic and hydrostatic bearings and therefore ignored in those cases. The two most important phenomena are briefly described below.

The term 'half-speed whirl' is applied to the effect experienced with the

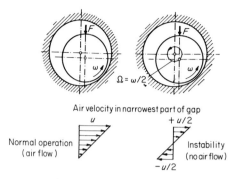

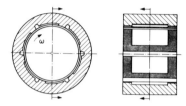

Fig. 4.94 Half-speed whirl on aerodynamic radial bearings (H. Drescher)

use of aerodynamic bearings, whereby the shaft undergoes an excitation at half of its rotating speed. This may be traced to a turbulence in the narrowest part of the air gap which interrupts the air flow in the circumferential direction. This causes a breakdown in the load-carrying capacity of the air in the narrowing gap on the pressurized side, as well as in the widening gap on the other side. The velocity distribution in the air gap and the displacement of the centre of the shaft are clarified in the upper right part of Fig. 4.94. Half-speed whirl may be avoided with the use of so-called 'dynamic air pressure' air bearings, as shown in the lower part of the diagram, which may be compared with hydrodynamic multi-bearing face bearings. However, due to the minute bearing clearances in aerodynamic bearings, their manufacture involves a very expensive production process.

Due to the compressibility of air, self-excited vibration effects may be experienced in aerostatic bearings, often known as 'air hammering', which are clearly audible and render them operationally useless. These phenomona—the detailed theory of which is not dealt with here—are particularly prominent when large bearing cells with few restrictor points are provided, as shown in the upper part of Fig. 4.95. The use of a large number of restrictor orifices prevents such vibrations. For this reason, bearings are produced today in a porous material, as shown in the lower part of the diagram, where a sintered sleeve is fitted between the air input and the bearing clearance gap. The porosity of the sleeve has the effect of many

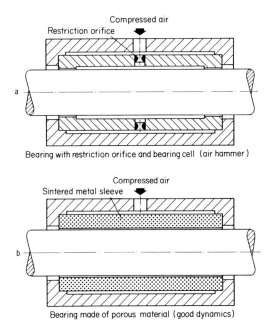

Fig. 4.95 Designs of aerostatic radial bearings
(J. Schmidt)

restrictor orifices or supply channels, with low tendencies to vibration of the air mass.[57]

4.3.3 Application examples

The application of air bearings in machine tools is predominantly limited to aerostatic bearings. Occasionally use is made of aerostatic guideways. Two examples are quoted below of aerostatic applications.

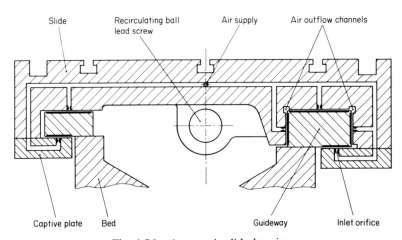

Fig. 4.96 Aerostatic slide bearing

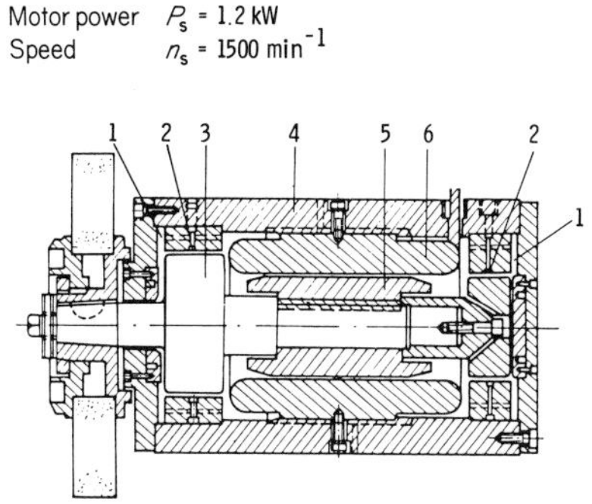

Motor power $P_s = 1.2$ kW
Speed $n_s = 1500$ min^{-1}

1 Thrust bearing 3 Shaft   5 Armature
2 Radial bearing 4 Housing   6 Stator

Fig. 4.97 Air bearing grinding-wheel spindle with built-in direct drive (A. Wiemer)

Figure 4.96 shows a cross-section of an aerostatic linear bearing for a table on a boring machine. The design consists of a load-bearing guide with a captive plate and a narrow guide. An advantage when compared with a hydrostatic guide is that there is no need to provide for the return flow of the lubrication medium and that there is no danger of contamination.

An example of an air bearing grinding-wheel spindle unit with direct-motor drive for large grinding wheels is shown in Fig. 4.97. The two end plates of the unit are used as bearing surfaces for the thrust bearings. Such units, which are particularly suited for precision grinding machines, offer the advantages of low bearing friction and consequential low heat generation, as well as a long life, accurate rotational running and simple construction.

4.4 Rolling guides and bearings

4.4.1 Rolling guides and guideways

Rolling linear guides and guideways are widely used in practice, alongside plain linear guides. The following advantages are obtained when compared with plain guides: light running forces due to rolling friction, no stick-slip, trouble-free installation and immediate availability due to standardization of the rolling elements.

The main disadvantage of this type of guide when compared with hydrostatic and hydrodynamic guides lies in the low damping effects in the direction normal to the movement.

4.4.1.1 Principles of construction

Figure 4.98 shows in its upper section a rolling guide using a roller chain. Generally, rolling elements travel half the distance moved by the slide. When roller chains are employed, their length must at least equal the sum of half the

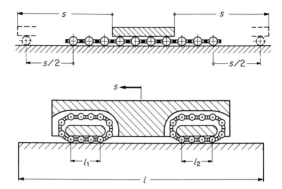

Fig. 4.98 Principles of construction for linear roller guides (Schneeberger)

slide travel plus its own length. Hence, for long movements rolling guideways in general employ recirculating roller elements, as shown in the lower half of Fig. 4.98. In this case the rollers run in an endless track, so that the movement is limited only by the length of the running surface.

For accurate operation of the complete rolling guideway, consideration must not only be given to the quality of both guiding surfaces but also to the dimensional and geometric accuracy of the rolling elements and their control in the cage. When cylindrical rollers are used, inaccurate cages as well as non-parallel guideway surfaces can lead to angular motion and side thrust of the rollers on the cage, which in turn will result in damage to the guiding surfaces and the cage itself, due to the friction generated. This problem is not encountered when balls are used as rolling elements, but such guides have the disadvantage of even lower stiffness and load-carrying capacity when compared with roller linear guides (point contact instead of line contact).

When rolling guides are in motion a reduction in stiffness is generally evident due to the constant change from n rolling elements to $n + 1$. For this reason it is of the utmost importance that the rolling elements should come into and roll out of the load-carrying zone of the guide as gently as possible without any jerky movement.

4.4.1.2 Types of design

Figure 4.99 illustrates the basic designs of rolling guideways. In the upper part, flat and vee guides are shown, whilst below these an angular roller chain and a ball-bearing guideway are presented. In these latter two types of guideway the cages carry the rolling elements at a 90° displacement to the applied force.

4.4.1.3 Pre-loading

Rolling guideways are pre-loaded as shown in the examples illustrated in Fig. 4.100 to improve stiffness and remove bearing play. The pre-load is applied

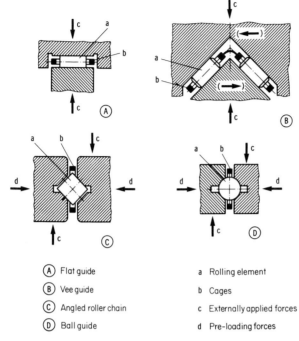

- Ⓐ Flat guide
- Ⓑ Vee guide
- Ⓒ Angled roller chain
- Ⓓ Ball guide

- a Rolling element
- b Cages
- c Externally applied forces
- d Pre-loading forces

Fig. 4.99 Various designs of rolling guideways (Schneeberger)

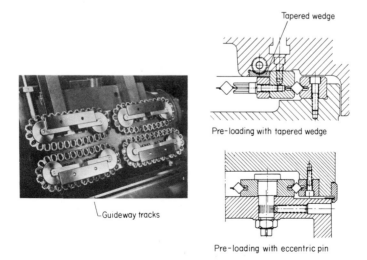

Guideway tracks

Pre-loading with tapered wedge

Pre-loading with eccentric pin

Fig. 4.100 Rolling guides with recirculating roller cages; various means of pre-loading of guideways (SKF)

by means of a tapered wedge (upper right in Fig. 4.100), set-screws or an eccentric pin (lower right).

4.4.2 Rolling cylindrical bearings

The most widely applied bearing principle for spindle bearings, drive shafts, as well as auxiliary functions, is that of rolling cylindrical bearings. The main reasons for their great popularity lie in the many advantageous characteristics of such bearings, particularly their international standardization, the comparatively simple design calculations and the wide selection available with the aid of catalogues and design indexes.

4.4.2.1 Survey of bearing designs

Figure 4.101 presents a diagrammatic summary of commonly used rolling bearings and their typifying characteristics. The various specifications demanded in use, such as load-carrying capacity, stiffness, axial misalignment and angular setting, higher-than-standard precision and speeds, are coped with by different bearing designs with varying degrees of effectiveness. Consequently, any one bearing type is more or less suitable for only one characteristic range of applications. As a generality it may be said, for example, that roller bearings have a greater load-carrying capacity than ball bearings of the same size. However, when it comes to combined radial and axial loading or to high running speeds, ball bearings will be more efficient. Hence, the selection

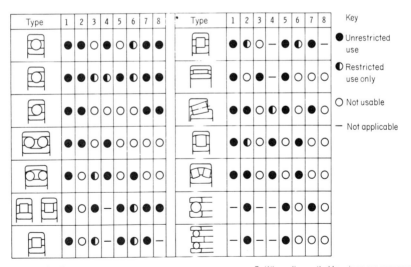

1 Radial loading
2 Axial loading
3 Linear self-adjustment of bearing with both rings fixed
4 Linear self-adjustment by angled seating in bore or outer ring
5 Where dismantled bearings are necessary
6 Self-adjustment for misalignments
7 Available in ultra-precision ranges
8 Where abnormal speeds are expected

Fig. 4.101 Commonly used rolling bearing and their features (FAG)

of a particular bearing type is governed by the specification of any one particular bearing duty (e.g. in terms of running speed, load-carrying capacity, direction of externally applied force, life, etc.).

4.4.2.2 Bearings for machine spindles and tolerances for the associated shafts and bearing seatings

Spingle bearings have a particularly difficult specification to fill because they are required to cater for high running accuracies even when under load and in a wide speed range. To achieve this, certain characteristics, e.g. high roundness accuracies and stiffness, will be called for.

For the reasons given above, special rolling bearings have been developed for machine-tool construction, distinguished by their high degree of accuracy, stiffness and low friction which avoid thermal expansion problems, among other special features. Figure 4.102 presents commonly used bearing types. Based upon the required precision of a machine, bearings are available with

Bearing Type	Precision designation	Mounting position	Tolerances		
			Tolerance range	Geometry	Alignment
	SP	Shaft		IT 2	IT 1
		Housing	K 5	IT 2	IT 2
	UP	Shaft		IT 1	IT 0
		Housing	K 4	IT 1	IT 0
	P 5	Shaft	js 5	IT 2	IT 2
		Housing	H 6	IT 4	IT 2
	P 4	Shaft	js 4	IT 1	IT 1
		Housing	H 5	IT 3	IT 1
	SP	Shaft	k 4 Fixed	IT 2	IT 2
			js 4 Adjustable	IT 2	IT 2
		Housing	K 5 Rigid bearing	IT 3	IT 2
			H 5 Free bearing	IT 3	IT 2
	P 5	Shaft	js 5	IT 2	IT 2
		Housing	JS 6 Rigid bearing	IT 3	IT 2
			H 6 Free bearing	IT 3	IT 2
	P 2	Shaft	js 3	IT 0	IT 0
		Housing	JS 4 Rigid bearing	IT 1	IT 0
			H 4 Free bearing	IT 1	IT 0
	SP	Shaft	h 5	IT 1	IT 1
		Housing	K 5 Rigid bearing	IT 2	IT 2
			G 6 Free bearing	IT 2	IT 2
	UP	Shaft	h 4	IT 0	IT 0
		Housing	K 4 Rigid bearing	IT 1	IT 0
			G 6 Free bearing	IT 1	IT 0

Fig. 4.102 Guidelines for the machining of bearing seatings and shaft mountings (FAG)

very close manufacturing tolerances, recognized by a designation stated after the bearing type, e.g. Sp = special precision, UP = ultra precision and normal tolerances P2 to P5.[58] Naturally, the use of high-precision bearings is only effective if the bearing seatings are machined to correspondingly tight tolerances, because the relatively thin bearing rings will readily distort as a result of any errors in the bearing seating or shaft diameter. The corresponding permissible geometric and alignment errors are governed by the precision of the bearing itself and are stated in terms of the ISO tolerance system (IT numbers) (BS 4500).

4.4.2.3 Bearing play

The terms 'bearing play' or 'bearing slackness' are applied to the relative movement of the bearing rings from one extreme position to the other when there is no applied load. Depending upon the direction of this displacement, differentiation may be made between radial and axial play. The play of a bearing is different when stationary compared with that when running (Fig. 4.103).

Normally, rolling bearings are produced with a particular degree of play, Δr_F. The amount of play left by the manufacturer places the bearing into one of five groups designated C1 to C5. Bearings with average play are in the groups C3 and C4.

After fitting the bearing rings into the housing and/or on the shaft, the play is reduced to Δr_E (Fig. 4.103, upper left). The degree of reduction in play is determined by the elastic expansion of the inner ring and the compression of the outer ring; these in turn are dependent upon the type of fit of the bearing rings on shaft and housing respectively, as well as the surface finishes of the contacting areas (cold or warm assembly).

Generally, when the bearing is running, a further change occurs in the bearing play, known as the running play Δr_r (on the right of Fig. 4.103). This

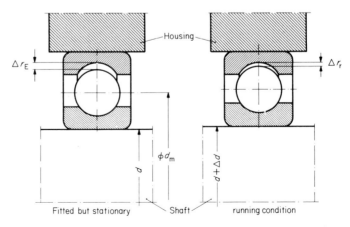

Fig. 4.103 Stationary and running play on a bearing (FAG)

may be caused by a considerable increase in the shaft temperature (due to localization of heat in the shaft) or, in the case of high-speed rotating bearings, by the expansion of the inner bearing ring through centrifugal force effects.

4.4.2.4 Resilience and pre-loading of radial bearings
The term 'bearing resilience' δ_r is applied to the opposing displacements of both bearing rings under the influence of an externally applied load. Figure 4.104 clarifies this phenomenon for different installation conditions of a radial bearing. On the left, a bearing with positive bearing play ($\Delta r > 0$) is shown.

When an external load F_r is applied the two bearing rings are displaced relative to each other due to the movement of the inner ring from the centre-line, by an amount $\Delta r/2$ plus the degree of elastic resilience $\delta_{e\,max}$ of the main load-bearing rolling element. When there is a positive bearing play the loaded zone ($-\psi_0$ to $+\psi_0$) of the bearing is less than 360°, so only a few rolling members support the externally applied load and the bearing resilience is correspondingly high.

When a bearing is pre-loaded a uniform elastic resilience of the rolling elements is obtained even before any external load is applied. Due to this initial deformation (flattening) of the rolling elements and because of the fact that all the rolling elements contribute to the support of the externally applied load (loaded zone = 360°), the ensuing bearing resilience is much less than that experienced by bearings having a positive play. A qualitative assessment

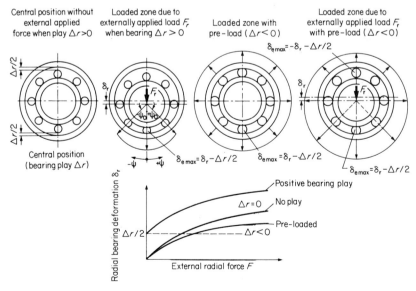

Fig. 4.104 Relationship between bearing play, loaded zone and bearing resilience (FAG/SKF)

of bearing resilience with respect to radial force for three different installation conditions is given by the curves shown in the lower part of Fig. 4.104.

It may also be noted that the resilience curves of a rolling bearing are not linear, because as the force rises the increasing deformation produces an improved stiffness.

The details of these relationships can be clarified in a stress diagram of a radial bearing, as in Fig. 4.105, where an analogy to the stress diagram of two springs is shown. When considering a bearing which has no play, then an externally applied radial force F_r produces a bearing resilience as indicated by the solid curve. If the bearing is pre-stressed, then this has the effect of superimposing a mirror-like curve of a second spring. In the diagram the bearing resilience due to an externally applied force F_r on a play-free bearing δ_r^* is indicated. If that bearing is pre-loaded so that $\Delta r = -\delta_r^*$, then the same externally applied force F_r will produce exactly half the previous bearing resilience. An increase of the bearing pre-load (e.g. to $\Delta r = -2\delta_r^*$) brings about a further reduction in the bearing resilience because use is made of the shape of the spring stiffness curve of the bearing.

For a quantitive assessment of radial-bearing resilience, nomograph–self-computing charts, based upon the calculation formulae by Lundberg and Stribeck[59] such as shown in Fig. 4.106,[60] are available for the commonly used spindle bearing in machine-tool construction. Such calculations assume that the bearing is without play when installed and that the following idealistic conditions apply:

(a) geometrically accurate rolling elements and bearing rings;
(b) geometrically accurate bearing seatings;
(c) both bearing rings secure;
(d) infinite stiffness in the bearing support components.

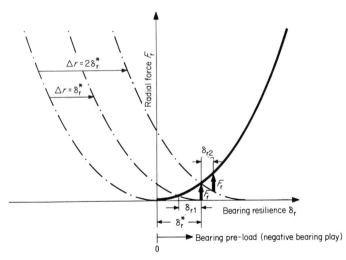

Fig. 4.105 Effect of pre-load on bearing resilience

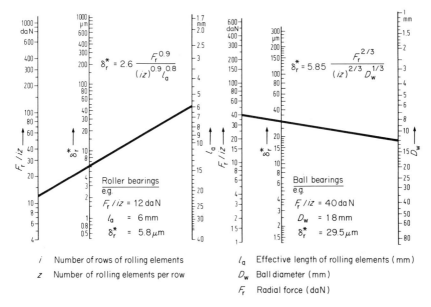

Fig. 4.106 Nomograph—self computing charts for determination of radial resilience of play-free rolling bearings (SKF)

In order to apply the charts, details of some bearing dimensions and specifications are required, i.e. i, z, l_a and D_w, which cannot always be found in catalogues and must be obtained from the bearing manufacturer for individual cases. With the aid of these charts, the resilience δ_r^* may be established in relation to the radial force F_r for a play-free installed bearing.[61]

From a resilience ratio β, which has no units, the resilience δ_r for a specific bearing play may be obtained based upon the resilience δ_r^*, i.e.:

$$\delta_r = \beta \delta_r^* \qquad (4.94)$$

The relationship between the relative bearing play and the resilience ratio β is given in Fig. 4.107 for roller and ball bearings. In order to facilitate an independent determination of the resilience ratio β for a given bearing, the abscissae of the curves shown have been normalized on the resilience of a play-free bearing (δ_r^*). It must be noted, therefore, that for a positive bearing play the bearing resilience due to the constant portion of half the bearing play $\Delta r/2$, as described previously, is already included.

The effect of the bearing-pre-load upon the bearing stiffness may be established directly from the reciprocal of the resilience ratio β for both definitions of stiffness given in section 2.4.1. In general, the following relationship is valid:

$$k(\Delta r) = k(\delta^*) \frac{1}{\beta} \qquad (4.95)$$

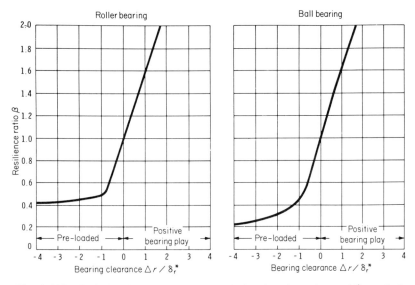

Fig. 4.107 Relationship between the relative bearing play $\Delta r/\delta_r^*$, and the deformation ratio (SKF)

It may be noted that approximately twice the load bearing capacity is obtained on roller bearings having a specific pre-load of $\Delta r/\delta_r^* = -1$ and that any further increase in the pre-load has an insignificant effect on the stiffness.

The position is somewhat different for ball bearings. In this case higher pre-stress values will bring further increases in the load-carrying capacity up to about five times the stiffness of a play-free bearing. However, when comparing the merits of both types of bearing, it must be remembered that the stiffness of roller bearings is considerably greater to start with than that of a ball bearing of the same size.

When considering the application of pre-loading for improved load-carrying capacity, there is an optimum value which should not be exceeded because the increase of stiffness obtained must be balanced against the increased rolling friction caused by the pre-load, the consequential increase in running temperature and the reduced bearing life. (The life of a rolling bearing is the running time that the bearing is able to give when correctly made, installed and maintained under prescribed loads.) An optimum pre-load value can be empirically stated as:

$$-0.8\delta_r^* > \Delta r_{opt} > -1.2\delta_r^* \qquad (4.96)$$

In this connection it must be borne in mind that this pre-load is to be operative in the working position of the bearing, i.e. when the bearing is running (running play). Hence, the pre-load must not be confused with the bearing's installed play or the play in the bearing after manufacture.

The effect of pre-load upon the bearing resilience for various loads is shown

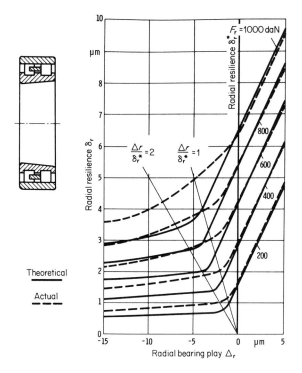

Fig. 4.108 Resilience of a double-row roller bearing NN3016K in relation to bearing play for different radial loads F_r (SKF)

in Fig. 4.108, using a double-row roller bearing as an example. The theoretical bearing resilience for different radial loads are plotted against the bearing pre-load and shown by the solid lines. Due to the effect of the stiffness of the bearing mounting components, the actual recorded resilience is in general somewhat greater, as shown by the dotted lines. From the diagrams it may be seen that the optimum bearing pre-load when related to the radial load lies in a range of -2 μm to -5 μm, i.e. $\Delta r/\delta_r^* \approx 1.6$.

4.4.2.5 Resilience and pre-loading of thrust bearings

Whereas in the case of radial bearings an individual bearing may be pre-loaded without the application of an external force, thrust bearings can only be pre-loaded with such a force (e.g. with another bearing or a spring). As may be seen from Fig. 4.109, rolling elements on bearings where the contact angle $\alpha > 0°$ are additionally subjected to a gyroscopic couple[62] (α is the angle between the direction in which the rolling element is loaded and the rotating plane of the bearing).

These gyroscopic couples are caused by the directional change of the axis of rotation of rolling elements and tend to turn them at right angles to the rolling

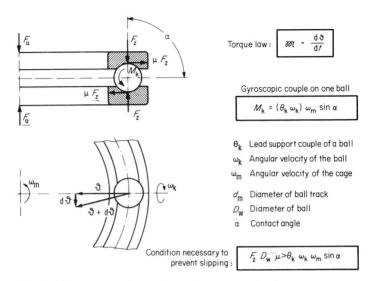

Fig. 4.109 Gyroscopic couple acting on a ball of a grooved ball thrust bearing

direction. The effect of this is—e.g. in the case of tapered roller bearings—that the bearing rings are forced apart due to the cant of the rollers. In the case of grooved-track ball thrust bearings, there is a danger that the gyroscopic couples will cause a rotation of the balls at right angles to their direction of rolling. It is for this reason that a minimum pre-load must be applied to prevent these gyroscopic motions, as otherwise slip will be caused between the rolling element and the running surface which leads to frictional wear. The exact conditions are shown in Fig. 4.109.

Firstly, the minimum pre-load must be such that the product $F_z D_w \mu$ is greater than the gyroscopic couple (F_z being the load per rolling element). Secondly, it must be remembered in the case of fast-rotating thrust bearings that the centrifugal forces acting on each rolling element tend to force the bearing rings apart, so that the line of contact is outwardly displaced. The minimum axial pre-load which takes account of both these effects is given by:

$$F_{a\,min} = M \left(\frac{n_{max}}{1000} \right)^2 \quad \text{with } F_{a\,min} \text{ in dekanewtons (daN)} \quad (4.97)$$

where M is the bearing constant given in the catalogue.

4.4.2.6 Comparison between radial and axial resilience curves for various bearings
The stiffness of rolling bearings is governed by their internal diameter, type of bearing and the pre-load applied (Fig. 4.106 to 4.108). Figure 4.110 shows a quantitative comparison of resilience curves of commercially available bearings when installed without play. It can be seen that a combination of double-

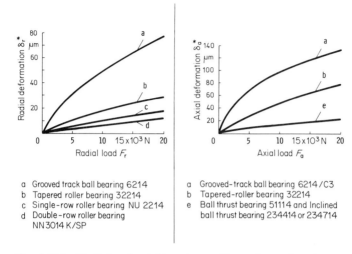

Fig. 4.110 Radial and axial bearing resilience for various types of bearing (SKF)

a Grooved track ball bearing 6214
b Tapered roller bearing 32214
c Single-row roller bearing NU 2214
d Double-row roller bearing NN 3014 K/SP

a Grooved-track ball bearing 6214/C3
b Tapered-roller bearing 32214
e Ball thrust bearing 51114 and Inclined ball thrust bearing 234414 or 234714

row roller bearings (d) with grooved-track ball thrust bearings (e) provide the best radial and axial stiffness.

However, such bearing types are not suitable for very fast running speeds (e.g. as required on grinding machines). Hence a compromise must often be accepted between adequate bearing stiffness and maximum rotational speed.

4.4.2.7 Cage slip on radial bearings
Normally the tangential frictional forces between the rolling elements and the bearing rings are greater than the forces necessary for the start and maintenance of the movement of the cage and its rolling members. These latter forces are largely governed by the hydrodynamic resistances to the movement of the lubricating medium. However, when there is only a light load on a bearing with positive play, the hydrodynamic forces may be greater than the tangential frictional forces and consequently the rolling contact is transformed into a sliding motion, i.e. the rolling elements and the cage move because of drag at a slower speed than their theoretical kinematic velocity. When the maximum slip has been reached, the irregular load peaks in the bearing ('sluggish' shaft motion in the bearing) will cause the slip to reduce again. This cycle of increase and decrease of slip will result in premature bearing breakdown due to rubbing and increase in wear of the rolling elements. Pre-loading may be applied to help reduce this effect.

4.4.2.8 Vibration excitations from rolling bearings
Any deviation from true roundness in the track or the rolling elements themselves (from errors) will lead to kinematic faults and consequently to irregular movements of the centre of the shaft (see also section 2.1.5.4). This, in turn,

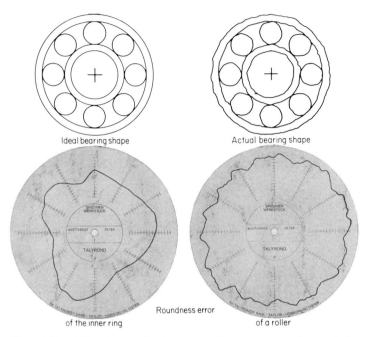

Fig. 4.111 Vibration excitations caused by errors in form of the bearing rings and rolling elements (SKF)

will excite vibrations. In general, long-pitched waviness in the track irregularities will cause inaccuracies in the rotation of the shaft (Fig. 4.111, left), whilst the shorter-pitched errors in the form of the rolling elements (Fig. 4.111, right) often govern the noise emitted from the bearing.

Apart from these roundness errors, the fundamental stiffness fluctuations between the 'apex positioning' and 'symmetrical positioning' of the rolling elements, as shown in Fig. 4.112 (occurring even in a perfectly round bearing), is also a source of excitation for vibrations. The frequency of these excitations (rolling frequency) may be obtained from:

$$f_E = 0.5n \left(1 - \frac{D_w}{d_m} \cos \alpha \right) \frac{z}{60} \quad \text{(Hz)} \tag{4.98}$$

The stiffness fluctuations will reduce as the number of rolling elements supporting the bearing load increases; hence pre-loading of the bearing is also advantageous from this point of view.

Moreover, the application of double-row rolling bearings, where the bearing elements are relatively displaced by half of one pitch, are of great advantage. For example, a 70% reduction in the vibration amplitude can be obtained from a double-row roller bearing compared with a comparable single-row roller bearing, with all other bearing conditions remaining equal (see the lower part of Fig. 4.112).

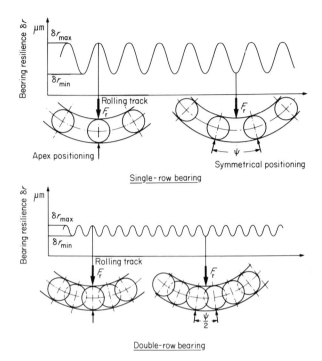

Fig. 4.112 Vibration excitations in a rolling bearing (SKF)

4.4.2.9 *Lubrication and temperature effects*

The primary function of a lubricant is to form a fluid film between the rolling element and the track, as well as the cage in the contact zone, thus preventing direct metal-to-metal contact and hence wear and corrosion of the bearing. In machine tools, rolling bearings are very frequently lubricated with oil. In this case the oil flow will also be expected to dissipate the heat generated in the bearing. A very close relationship exists between the quantity of oil employed and the bearing temperature as may, for example, be seen from the experimental results for power loss and bearing temperature plotted against oil quantity in Fig. 4.113.[63] When the oil quantity Q_S is insufficient, the lubrication of the bearing is inadequate, causing the power loss and the bearing temperature to be unnecessarily high. As the quantity of oil is increased the power loss and temperature decrease, until a minimum is reached when the lubrication is adequate. Any further increase in the oil supply causes an increased power loss (splash losses) and the temperature will also tend to rise at first due to these splash losses. However, the temperature drops again when the heat dissipated by an increased quantity of oil is greater than the heat generated by the splash losses (cooling effect).

The effects described above occur in an oil-flow lubrication system of the type shown on the left of Fig. 4.114. At higher rotational speeds, this method

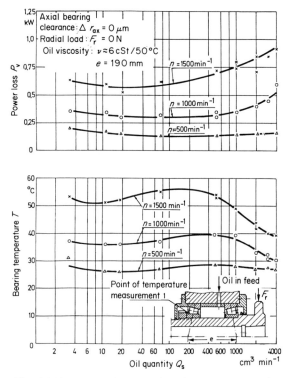

Fig. 4.113 Power loss and bearing temperature in relation to the quantity of oil supplied

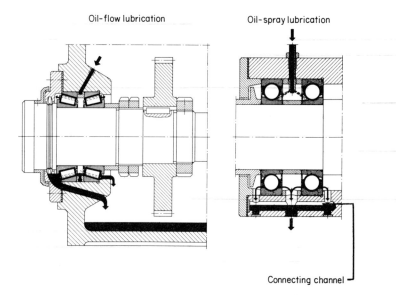

Fig. 4.114 Oil lubrication systems (FAG)

of lubrication is unsuitable. In these cases an oil-spray lubrication may be used, as shown on the right of Fig. 4.114, where a smaller quantity of oil is sprayed with compressed air directly into the contact zones. When the bearing position is correctly chosen for a given application and the bearing carefully mounted, this type of lubrication system permits the use of the smallest quantities of lubricant coupled with reliable lubrication of the rolling bearing.

Grease lubrication should only be applied when the rotational speeds are low and when no heavy bearing loads may be expected. Almost all modern electric motors are fitted with grease-lubricated bearings because they require little maintenance. In addition, grease-lubricated sealed bearings are frequently used for vertical spindle bearings owing to the simplicity of the construction.

4.4.2.10 Characteristics of rolling bearings compared with other bearings

Figure 4.115 is a comparative presentation of the most important characteristics of rolling, hydrodynamic and hydrostatic bearings. Detailed consideration was given to the characteristics of the latter two types in previous sections (see sections 4.1 and 4.2) and further discussion is not necessary here.

The following advantages may be claimed for the use of rolling bearings:

(a) interchangeability due to standardization;
(b) high level of standardization;
(c) availability of catalogues and design indexes to aid selection;
(d) small-width space requirements;
(e) high load-carrying capacity, even when the rotational speed $n = 0$, because, unlike in a plain bearing, a lubrication film need not first be established ($n > 0$);

Characteristic	Hydrodynamic bearing	Rolling bearing	Hydrostatic bearing
Damping	●	○	●
Running accuracy	●	◐	●
Speed range	○	◐	●
Wear resistance	◐	◐	●
Power loss	●	○	◐
Installation costs	◐	○	●
Cooling capacity	◐	◐	●
Reliability	●	●	◐

Evaluation of characteristics:
● high   ◐ medium   ○ low

Fig. 4.115 Characteristics of commonly used spindle bearings

(f) high fluctuations of loading and rotational speed are acceptable;
(g) in some circumstances radial bearings can also act as thrust bearings;
(h) the rolling action leads to only minimal friction ($\mu = 0.0015$);
(i) low power loss and bearing temperatures;
(j) maintenance is less frequently required when correctly installed;
(k) running accuracy and stiffness may be improved by pre-loading.

Among disadvantages brought about by the use of rolling bearings, the following may be included:

(a) lubrication and cooling problems at high rotational speeds;
(b) reduced load-carrying capacity at high rotation speeds;
(c) large radial space requirements;
(d) heavy;
(e) expensive when special bearings are required:
(f) limited life if overloaded (pitting);
(g) allowance must be made for the centrifugal force effect in the case of thrust bearings as well as inclined and taper roller bearings;
(h) vibration excitations and noise generation.

4.4.3 Spindle bearing units employing rolling bearings for machine-tool construction

4.4.3.1 Specifications and design principles

The following may be regarded as the main functions of main and working spindles in metal-cutting machine tools:

(a) the guiding of the tool and/or work at the cutting point with adequate kinematic accuracy;
(b) the absorption of externally applied forces such as the weight of the workpiece and cutting forces with minimum static, dynamic and thermal distortions.

The dimensional accuracy and surface finish of the work being machined, as well as the rate of metal removal of a machine tool, are, among other factors, directly governed by the static, dynamic and thermal behaviour of the spindle bearing unit. Hence special attention must be paid to the design of this part of the machine's construction. In general terms, there are three different basic design concepts which are widely applied to the construction of machine-tool spindle bearings:

(a) tapered roller bearings in the lower end of rotational speed range, good stiffness (turning, milling);
(b) double-row parallel roller bearings in the medium rotational speed range, very good stiffness (turning, milling, drilling, and grinding);

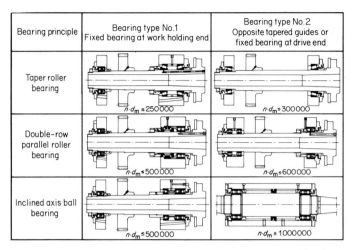

Fig. 4.116 Commonly used spindle bearing designs (SKF)

(c) inclined angular contact ball bearings in the upper rotational speed range, poor stiffness (grinding, turning and milling of non-ferrous metals).

Figure 4.116 provides a summary of six typical spindle bearing designs. The stiffness of the bearings illustrated decreases from the upper diagram to the lower, whilst the permissible speed range increases. The alternative bearing type 2 is more flexible, but may be used for a higher speed range than type 1. The most important characteristics of the various spindle designs are summarized in Table 4.5.

When rolling bearings are fitted in pairs and the contact angle $\alpha \neq 0°$ (e.g. tapered roller bearings, angular contact ball bearings), then one of three fundamentally different mounting arrangements of the bearings may be chosen, as shown in Fig. 4.117 using an angular contact ball bearing as an example. The 'O' arrangement shown on the left has the advantage of providing a good end restraint for the shaft due to the line of action for the force transmission of the balls, and is, therefore, particularly well suited for the absorption of cantilever loads. Furthermore, this arrangement is able to deal with axial loads in both directions. The 'X' arrangement in the centre of the figure can also cope with axial loads in both directions, in contrast to the 'tandem' arrangement shown in the right, which is capable of supporting greater axial loads, albeit only in one direction.

When angular contact ball bearings or tapered roller bearings are mounted in pairs and with opposite directions of the axes of the rolling elements, then the radial load is evenly distributed between them. Such double bearings are supplied by the manufacturer in pairs for particular applications, so that a prescribed bearing play exists without the need for the use of shims.

Figure 4.118 shows the spindle bearings of a machining centre designed to cope with high rotational speeds. The front bearing consists of a pair of

Table 4.5 Comparison of advantages and disadvantages of bearing designs shown in Fig. 4.116

	Bearing type 1	Bearing type 2
Tapered roller bearing	Wide support width of rollers provides good static stiffness	Simple, economic
	Good running accuracy when bearing has no play; only low rotational speeds possible	Danger of seizure or excessive play
		No thermally influenced change in bearing play if the axes of the rollers intersect on a single point on the spindle axis (see Fig. 4.129)
Double-row parallel roller bearing	Parallel roller bearing for radial loads; provides very good stiffness	Good radial stiffness
	Axial control with angular contact ball thrust bearing	Suitable for higher rotational speed, because the angular contact ball thrust bearing is replaced with a grooved-track ball bearing
	Improvement of total stiffness of unit by optimum adjustment of bearing play	Spindle expands forward if temperature rises
		Poor axial stiffness
Angular contact ball bearing	Alternative to tapered roller bearing for higher rotational speeds	Each bearing station in tandem layout, total system in 'O' layout
	Poor radial stiffness	Medium radial stiffness
	Stiffness improvement due to series mounting of a number of matched bearings	Suitable for very high rotational speeds (grinding)
	Low friction forces	Saucer spring unit improves running accuracy by pre-loading the bearing

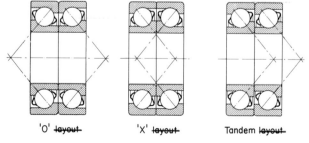

'O' layout   'X' layout   Tandem layout

Fig. 4.117 Various layouts of paired inclined axis ball bearings (FAG)

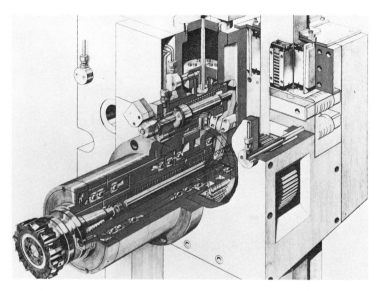

Fig. 4.118 Spindle bearings for a machining centre (Deckel)

precision angular contact ball bearings arranged in a tandem layout, as well as a third similar bearing which is mounted in an 'O' layout relative to the other two. The rear bearing is made of just two precision angular contact ball bearings, fitted in an 'O' layout.

4.4.3.2 Static behaviour
The accuracy of the work produced by a main spindle unit is governed by the total deflection of the spindle at the point-of-force application in the direction of the y axis.[61] Such a deflection is built up from a number of contributory elements, as may be seen in Fig. 4.119:

Spindle flexure element:

$$y_{Sp} = \frac{Fba^2}{3JE}\left(1 + \frac{a}{b}\right) \quad (J = \text{constant over the length}) \quad (4.99)$$

Bearing flexure element:

$$y_L = \frac{F}{b^2}\left[\frac{(a+b)^2}{k_A} + \frac{a^2}{k_B}\right] \quad (4.100)$$

The contribution to the total by the deflection of the frame (bearing housing) y_k is difficult to calculate; the finite element method may be applied (see section 2.6.2).

The total flexure is given by:

$$y = y_{Sp} + y_L + y_k \quad (4.101)$$

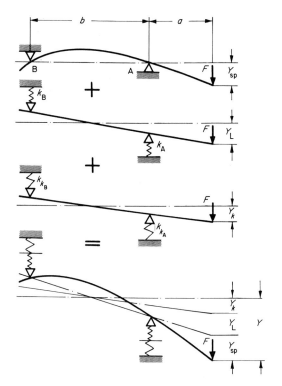

Fig. 4.119 Deflections of a main spindle

Because of the series connection of the individual contributions to the total flexure, the total flexibility at the point-of-force application is given by:

$$\frac{1}{k} = \frac{1}{k_{Sp}} + \frac{1}{k_L} + \frac{1}{k_k} = \frac{y_{Sp}}{F} + \frac{y_L}{F} + \frac{y_k}{F} \qquad (4.102)$$

The total flexibility of a spindle–bearing system is dependent upon the combined effect of the bearing stiffness, the centre distance between the bearings b, the length of the cantilever and the geometry of the spindle.

Figure 4.120 presents the contribution of the stiffness k_A of the front bearing position to the total flexibility of a particular spindle–bearing system. It may be seen that a bearing stiffness $k_A > 750$ N μm^{-1} has almost no influence on the total stiffness of the system, because the flexibility of the spindle is then an ~~overriding~~ factor.

In this type of two-bearing system, a distance between the bearings can be established which will give a minimum deflection at the point-of-force application, given that all other parameters are unchanged (static optimum bearing centres).[64] If the equation for the total flexure, $y = y_{Sp} + y_L + y_k$, is differentiated with respect to the bearing-centre distance b and equated to zero (maxima–minima), then a cubic expression is obtained for b, the solution for which

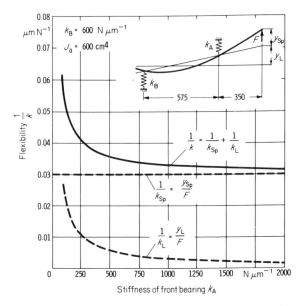

Fig. 4.120 Flexibility of a system in relation to the stiffness of the front bearing (bearing contributions shown separately)

may be obtained from the nomogram in Fig. 4.121. The value of a in the expression for y_2 must be expressed in centimetres. With the increasing application of mini-computers in design, such deformation calculations and optimizations are mostly carried out with the aid of computer programs.[61] *increasingly*

Quite often the bearing-centre distance must be increased due to other design considerations (e.g. the drive). In such cases a third bearing may be used (step bearing) which will improve the static behaviour. To this end, another option available is to use a drive which does not apply a shear load, thus reducing the effective load fluctuation on the bearing in service.

Figure 4.122 shows as an example the bearings of a vertical quill and spindle. The design incorporates the following features:

(a) double-row parallel roller bearings as main bearings;
(b) grooved-track ball thrust bearing used in conjunction with an angular contact ball bearing;
(c) torque is applied without shear stress by means of a bevel gear wheel and a splined shaft;
(d) the quill is moved axially by means of a rack and pinion;
(e) the quill can be clamped by means of a split plain guide;
(f) bearings are grease lubricated, because high rotational speeds are not required and oil lubrication on a vertical spindle is more expensive due to the need for seals.

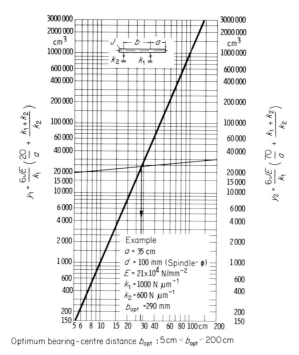

Fig. 4.121 Nomogram for the determination of the static optimum bearing-centre distance

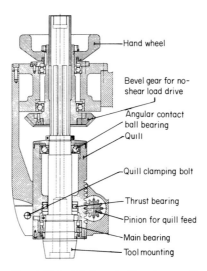

Fig. 4.122 Main spindle of a small vertical boring machine (SKF)

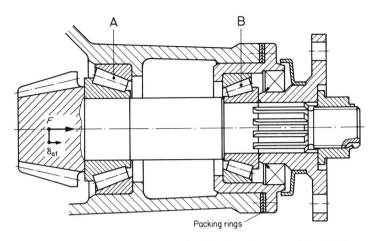

Fig. 4.123 Pre-loaded tapered roller bearing assembly (SKF)

Apart from the radial stiffness of the bearings mentioned earlier and their influence on the total deflections, the axial stiffness of a spindle–bearing system must also be considered. In general, the axial stiffness of the spindle is considerably greater than that of the bearing. To achieve very good stiffness, opposing pre-loaded taper roller bearings are highly suitable, as shown on the drive shaft in Fig. 4.123.

The effect of pre-loading on the elastic deformations caused by the applied load may be clarified with the aid of a load diagram. Figure 4.124 presents the flexure curves of the pre-loaded bearings and the mountings as shown in Fig.

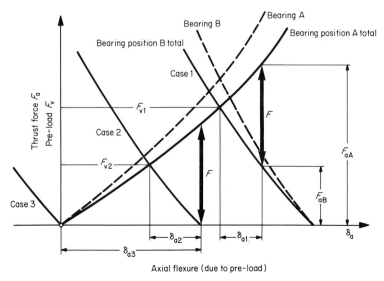

Fig. 4.124 Relationship between pre-load and flexure (SKF)

4.123. The curve which rises from left to right represents the components on which the machining forces add to the total load; components on which the machining forces act in an opposite way to the pre-load are represented by the curves which rise from right to left. It may be seen from Fig. 4.124 that the bearing mounting components have a softening effect on the system (the broken line shows the bearing only; the solid line, the bearing and mounting components).

Case 1: Bearing A is pre-loaded with F_{v1} against bearing B. Additional thrust load F (on pinion) produces deflection δ_{a1}.
Case 2: $F_{v2} < F_{v1}$; F produces flexure δ_{a2}. Bearing B becomes just about play-free.
Case 3: Bearing position A, B has no pre-load (but it has no play). F produces a flexure δ_{a3} ($\delta_{a3} > \delta_{a2} > \delta_{a1}$). Bearing B is completely without loading and play and backlash is introduced, i.e. the axial stiffness is considerably reduced when compared with case 1.

4.4.3.3 Dynamic behaviour
The dynamic characteristics of the spindle bearing system have a considerable influence upon the geometric accuracy and surface finish of work being machined, as well as on the vibration-free useful cutting power available. The criteria for dynamic behaviour are: resonance frequency, resonance amplitude, mode shape and the damping system (see Volume 4). The damping system consists of the damping properties of the spindle material and—if the effect of the mounting components is ignored—the damping in the bearing. A number of experiments have shown that bearing damping is mainly the result of viscous (velocity proportional) damping, which is known as the bearing damping coefficient c (in newton-seconds per metre) for which at this time only a few numerical values are available.

Figure 4.125 shows the bearing stiffness and bearing damping of an oil-lubricated double-row parallel roller bearing with respect to rotating speed as observed on a test bed under laboratory conditions. The bearing stiffness shown in the upper graph increases with the rotating speed due to the reduced bearing play induced by the running conditions. The upper curve is for a play-free mounted bearing, while the lower curve refers to a bearing with positive play ($\Delta r = 3\delta_r^*$).

The bearing damping shown in the lower graph is approximately 2000 N s m^{-1} when the spindle is at rest. When the spindle is set in motion there is at first a steep increase in the damping constant, but this cannot be clearly presented on the graph. After a speed of approximately 100 rev min^{-1} has been reached, a reduction in the damping constant is noted down to a minimum of about 2000 N s m^{-1}, but further increases in rotational speed cause the curve to rise again. The curves appear to follow the same pattern for both the play-free and the positive-play bearings.

The experimental results quoted above are early conclusions obtained from

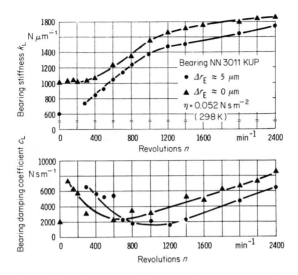

Fig. 4.125 Bearing stiffness and damping with respect to rotating speed on a double-row parallel roller bearing

a systematic research programme for rolling bearings. It is not yet clear to what extent the findings are representative for this specialized type of bearing.

Figure 4.126 presents the resonance curves of the displacement at the point of loading for the analogous model of the spindle–bearing system shown. The resonance curves were calculated on the assumption that the bearing damping coefficients c are not influenced by the bearing stiffness. In principle, the main

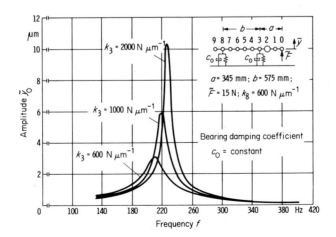

Fig. 4.126 Amplitude characteristics with respect to bearing stiffness

interest is to obtain the greatest possible static stiffness at the point of loading. As may be seen from Fig. 4.125, the resonance peak of the flexibility of the spindle–bearing system rises with the bearing stiffness. The reason for this apparently contradictory situation lies in the fact that when the bearing stiffness increases, no significant movement can occur within the bearing, which is the main source of damping for the complete system.

At a high bearing stiffness ($k_3 = 2000$ N μm^{-1}) only the damping characteristics of the spindle material remains. The natural bending mode of the spindle has its nodes in the bearing. Consequently, there remains no movement within the bearing.

As the dynamic characteristics of a given rolling bearing can only be changed within narrow limits (e.g. by the degree of pre-load), the dynamic behaviour of a spindle–bearing system can in the main only be influenced by the arrangement of the selected bearings and by the spindle geometry.

To this end, the designer should strive to achieve the shortest overhang possible (low mass with high spring stiffness) and as sufficiently large a spindle diameter coupled with the shortest bearing-centre distance as practical. Frequently these criteria cannot be fully satisfied due to other design constraints (spindle drives, mounting of main drive, connection of work-holding devices). In such cases, the following possibilities may be considered for further improvement of dynamic behaviour:

(a) restraint of spindle at the front bearing position (tool) by the use of several pre-loaded bearings (Fig. 4.127);
(b) damping of the bending vibrations of the spindle with the use of a damping element (Fig. 4.128).

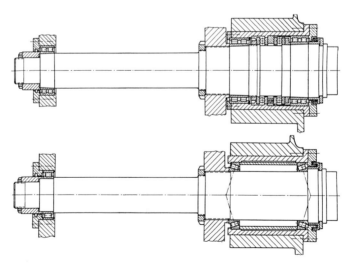

Fig. 4.127 Two main spindle-bearing arrangements (three-bearing system) (SKF)

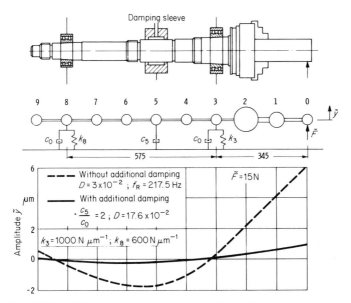

Fig. 4.128 Vibration modes of a spindle with and without a damping sleeve

The main spindle systems shown in Fig. 4.127 obtain an additional static stiffness through the use of a second main bearing (restraining effect); consequently, there is an improvement in the dynamic behaviour through the retention of the mass distribution pattern. Moreover, the second main bearing itself contributes to the damping of the system. It must be borne in mind that an increase in the production costs will occur and that an additional heat source is introduced which may require the provision of additional cooling.

Figure 4.128 depicts the effect of an additional damping element on the bending curve. The damping unit consists of a sleeve with a large oil gap height (e.g. 400 μm) and is placed at the mass point 5 in the analogous diagram. Due to the large fluid gap, the sleeve has no radial static stiffness but only exhibits a damping effect (squeeze-film damping).[65] In the example shown, the damping sleeve increases the damping of the system from $D = 3 \times 10^{-2}$ to $D = 17.6 \times 10^{-2}$; the resonance amplitude is reduced from 6 to 1 μm (the excitation force $F = 15$ N).

The damping sleeve may also be made in a tapered form, so that the fluid-film gap may be varied by axial movement on a correspondingly shaped spindle and an optimum position may be found.

4.4.3.4 *Thermal behaviour*

Apart from the static and dynamic behaviour of spindle–bearing systems, it is also important to consider thermal behaviour, because the heat generated due to the power losses at the rolling bearings may lead to considerable distortions of the headstock and consequently thermal deformations of the spindle (see section 2.6). Such radial and axial deformations have a direct influence on the accuracy of the work being produced. For this reason, thermoelastic displacements in the direction of the cutting point should be as small as possible. The axial spindle expansion may have an almost negligible effect on the accuracy of the work produced if the thrust bearing is placed close to the main bearing. In the case of widely spaced pre-loaded thrust bearings, unacceptable changes in the pre-load may occur due to thermally activated movements of the bearing centres. For this reason, such a design should be avoided whenever possible. In the case of taper roller bearings, the bearing positions should be chosen such that the axes of the rolling elements intersect at a common point on the spindle axis in order to reduce thermally activated changes in the pre-load (Fig. 4.129). In that case, an equal change in the radial and axial temperatures will cause any variation in the axial pre-load to be cancelled out by a compensating change in the radial pre-load.

The radial displacements may be considered in the design stage by the calculation of the effect of different variables (e.g. with the aid of the finite element method within given boundary conditions) which will examine the

Fig. 4.129 Thermosymmetric design of taper roller bearings

thermal deformations of different spindle-casing designs (for the establishment of a thermosymmetrical design, see section 2.6).

Finally, in the case where machine tools exhibit thermal displacements in practice, it is possible to compensate for these automatically by a corresponding correction in the tool feed (see Volume 3).

4.4.4 Recirculating ball spindles and nuts

The development of NC machine tools (see Volume 3) demanded among other things the provision of feed and guiding systems free from stick-slip and backlash. This required the availability of low-friction driving and guiding elements. Rolling guideways and recirculating–ball screw systems are the outcome of efforts in that direction. A recirculating–ball screw system may be regarded as a rolling guide on a shaft. Like all other screw–nut systems, it serves to transform rotary motion into linear motion, or vice versa. The main areas of application are for feed mechanisms and in measuring machines.

The principal positive features of a recirculating ball screw are:

(a) very high mechanical efficiency due to rolling friction (up to 95%);
(b) no stick-slip;
(c) when correctly designed almost no wear and hence a very long life;
(d) able to be pre-loaded (no play);
(e) adequate stiffness.

The only disadvantage which must be mentioned is their low damping property.

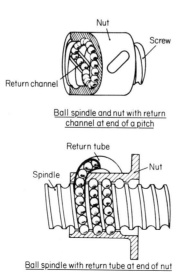

Fig. 4.130 Recirculating ball screw and nut systems (Warner)

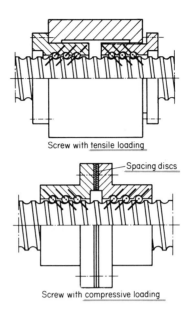

Fig. 4.131 Methods of pre-loading ball screw and nut systems (Warner)

As the balls rotate in the guiding grooves of the screw and nut, they are subjected to a tangential or circumferential movement. Hence it is necessary to provide for the return of the balls into the system. Figure 4.130 shows in the upper diagram a screw with a ball-return track at the end of each thread pitch. The advantage of this design lies in its small physical size. The unfavourable angles of the ball inlet and exit bring about the disadvantage of uneven rolling effects.

In the lower part of the diagram a ball screw is shown on which the ball return is effected over the full length of the nut by means of a return tube. By suitably shaping the tube, the balls leave and return to the load-bearing part of the nut tangentially, resulting in an even and shock-free running condition and permitting high rotational speeds to be used. An important disadvantage of this design is the ease with which the return tube may be damaged; this would hinder the ball motion and thus lead to further damage of the screw–nut assembly.

If a high degree of stiffness and/or freedom from backlash is required, then the ball screw and nut system must be pre-loaded. To this end the nut must be in two parts (double nut). Figure 4.131 shows such a system. In the upper part a double nut with an external thread is shown. The pre-load is applied by turning one half of the nut; subsequent keying then prevents it turning in the opposite direction. In the lower half of the diagram calibrated spacing discs are placed between the nut halves to produce the pre-load.

Whether the ball screw is to be pre-loaded with a tensile or compressive load is governed by the direction of the largest externally applied load. To maintain a good stiffness, care must be taken to ensure that a minimum pre-load remains even after the application of an externally applied load (see section 4.4.2.4).

5

MAIN DRIVES

5.1 Motors

The sections dealing with main drive units begin with motors which may be classified into two main groups as shown in Fig. 5.1: 'electrical main drives' and 'hydraulic main drives'. Both types are used in machine tools for the provision of the main working motions. These include the movement of the main spindle in lathes, milling machines, drilling machines, grinding machines or saws, the table movement on a planing machine and the ram motion on presses and slotting machines. The choice between electric or hydraulic motors is governed by the particular function expected from the drive. The hydraulic drive offers a reduction in weight for a given power and a better acceleration due to the smaller moment of inertia of its mass; the electric drive features a longer life, higher efficiency and reduced heat generation.

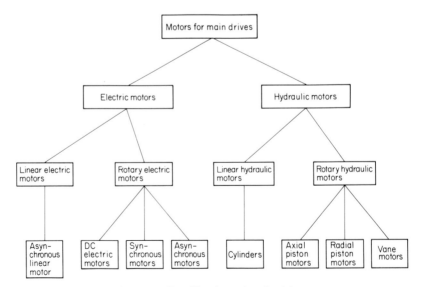

Fig. 5.1 Classification of main drives

	Main drives	Secondary drives	Auxiliary drives
Speed	Fine stepped	Very fine stepped	Constant
Speed range	Wide	Very wide	—
Speed regulation	Good	Very good	Poor
Even running	Good	Very good	Reasonable
Dynamic behaviour	Medium	Very good	Low

Maximum performance and torque according to the needs of the machine.

Fig. 5.2   Requirements in electric machine-tool drives

The desirable features for main drives with regard to their behaviour in service and output capacity may be seen in Fig. 5.2. For comparative purposes, the requirements in secondary drives and auxiliary drives are also shown. Secondary drives (Volume 3) provide feed motions for tables, slides and spindles, whilst auxiliary drives are used for cooling-fluid pumps, fans and other units. When determining the spindle-speed range in, for example, a lathe, the relationship for a given cutting velocity $n = v/d\pi$ must be borne in mind, as well as the diameter range of the potential work of the machine. The speed regulation of the drive must be such that the cutting velocity remains approximately constant under changing load conditions.

In the following sections, a number of types of motor are introduced, their basic construction and working principles described, their limitations in performance indicated and attention is drawn to the possibilities of adjusting their rotational speeds.

5.1.1   Electrical machines

5.1.1.1   Direct current (DC) machines

Separately excited DC motors are often used for main drives of machine tools because under controlled conditions they are able to provide a low-speed regulation, i.e. high torsional stiffness under load, and their rotational speed may be varied over a wide range. An electrical energy source which can provide a variable DC current will be required for the drive of a DC machine.

Construction and working principles. The general design and circuit diagram of a separately excited DC machine are shown in Fig. 5.3. An excitation current I_f produces a magnetic flux ϕ in the stator which is passed into the rotor across an air gap. The rotor current flows through the rotor windings, which are represented in the diagram as a single coil perpendicular to the plane of the drawing. In accordance with the principle that a magnetic field produces a force in a current-carrying conductor, a torque is generated in the direction indicated. In order that the direction of the torque (rotation)

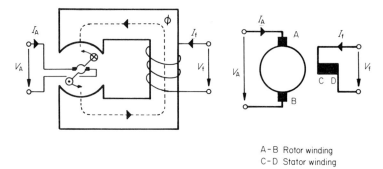

A-B Rotor winding
C-D Stator winding

Fig. 5.3 General design and circuit diagram of a separately excited DC motor

remains constant, the rotor current must be reversed after every half-revolution of the rotor (commutated). This is achieved with the aid of the commutator segments on the shaft which carry the rotor current through carbon brushes into the rotor windings with changing polarity. For the provision of an even torque, the windings on the stator and rotor of a DC machine are distributed around the circumference.

Another physical effect is of importance for the functioning of a DC machine. Due to the motion of the rotor windings in the stator field, a reverse voltage (back e.m.f.) is induced into the rotor circuit, which increases in proportion to the rotational speed and is of opposite polarity to the rotor current. The result of this effect on the performance of the machine is explained in the following section.

Basic and operational equations. The basic equations for a DC motor are:

$$V_A = k_1 \phi n + R_A I_A \tag{5.1}$$

$$T = k_2 \phi I_A = \frac{k_1}{2\pi} \phi I_A \tag{5.2}$$

In the above, V_A is the rotor voltage, R_A the ohmic internal resistance of the rotor, I_A the rotor current and ϕ the magnetic flux produced by the exciter windings. The motor constants k_1 and k_2 are dependent upon the construction, size and winding arrangements of a given machine.

When starting ($n = 0$), the induced back e.m.f. V_i is given by:

$$\text{For } n = 0: \quad V_i = k_1 \phi n = 0 \tag{5.3}$$

The maximum rotor current $I_{A\,max}$ will be limited in accordance with equation (5.1) only by the internal resistance of the rotor, i.e.:

$$\text{For } n = 0: \quad I_{A\,max} = \frac{V_A}{R_A} \tag{5.4}$$

Correspondingly, the motor torque available for the acceleration of the rotor and connected load will be:

$$T = k_2 \phi I_{A\,max} = k_2 \phi \frac{V_A}{R_A} \tag{5.5}$$

In proportion to the increasing rotational speed the induced back e.m.f. will increase according to equation (5.3) until its magnitude equals the value of the rotor input voltage, which is reduced by the load-dependent rotor voltage drop $I_A R_A$.

Thus, for 'no-load' running of the motor we have, from equation (5.2):

$$\text{For} \quad T = 0: \quad I_A = 0$$

When the rotor current $I_A = 0$, then from equation (5.1) we have, for the 'no-load' rotational speed:

$$\text{For} \quad V_A = V_i: \quad n_0 = \frac{1}{k_1 \phi} V_A \tag{5.6}$$

When loaded, we obtain, from equations (5.1) and (5.2), the rotational speed in accordance with the loading condition:

$$n = \frac{V_A}{k_1 \phi} - \frac{R_A}{k_1 k_2 \phi^2} T \tag{5.7}$$

This relationship is illustrated in the graphs for rotational speed plotted against torque in Fig. 5.4. The curves represent three different rotor voltages (V_{A1-3}). In each case, the rotational speed reduces from the 'no-load' condition as the load on the motor rises.

Speed control and load limitations. The speed of a DC machine may be controlled by the rotor (armature) voltage V_A (armature or primary control) or by the magnetic flux ϕ (field or secondary control). According to equation (5.7), an increase in the armature voltage increases the rotational speed. In contrast, an increase in the magnetic flux ϕ brings about a reduced speed. In both speed-control methods different load limitations apply which are clarified in Fig. 5.5. In the upper part of the diagram the two regulating factors—armature voltage and magnetic flux—are plotted against rotational speed; in the lower graph, the appropriate family of characteristics $T_{(n)}$ are shown, as well as the limiting curves for the load torque T_{max} and the permissible rotor current $I_{A\,max}$. In addition, the maximum useful output power at the motor shaft P_{mech} is shown, as well as the electrical input power for the system P_{max}. The maximum permissible continuous rotor current (continuous applied voltage) which is limited by the thermal loading capacity of the windings is assumed to be constant in both cases:

$$I_{A\,max} = \text{constant} \neq f(n) \tag{5.8}$$

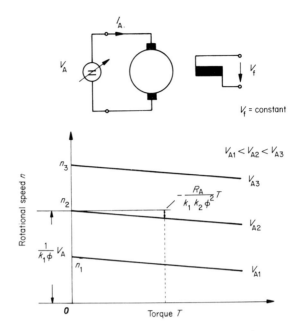

Fig. 5.4 Rotational speed–torque curves dependent upon rotor voltage

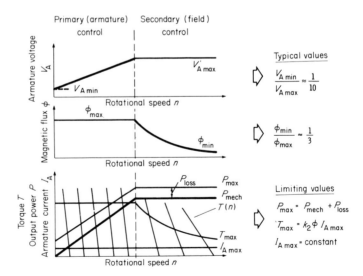

Fig. 5.5 Primary (armature) and secondary (field) speed control range

Short periods of overload are permissible depending on the heat capacity of the armature and the provision for heat dissipation.

Armature control. Speed control by armature voltage is suitable for low to medium speed ranges coupled with heavy motor torque loading. The maximum excitation provides the lowest, i.e. best possible, speed regulation:

$$V_{A\,min} \leq V_A \leq V_{A\,max} \tag{5.9}$$

$$\phi = \phi_{max} = \text{constant} \tag{5.10}$$

From equations (5.1) and (5.7) we obtain the relationship between rotational speed and armature voltage:

$$n = \frac{V_A}{k_1 \phi_{max}} - \frac{R_A I_A}{k_1 \phi_{max}} = \frac{V_A}{k_1 \phi_{max}} - \frac{R_A T}{k_1 k_2 \phi_{max}^2} \tag{5.11}$$

As the magnetic flux is constant in the armature-control method range (equation 5.10), there exists a linear relationship between the speed and the armature voltage. The speed reduces by a value dependent upon the applied torque load:

$$n \propto V_A - CT \tag{5.12}$$

The permissible continuous applied loading T_{max} may be obtained directly from equation (5.2):

$$T_{max} = k_2 \phi_{max} I_{A\,max} = \text{constant} \tag{5.13}$$

As a result of the conditions arising from equations (5.8) and (5.10) a constant maximum torque load can be obtained.

To obtain the maximum continuous electrical power input, the basic equation (5.1) is multiplied by $I_{A\,max}$:

$$P_{mad} = V_A I_{A\,max} = k_1 \phi_{max} n I_{A\,max} + R_A I_{A\,max}^2 \tag{5.14}$$

This is in accordance with the power balance:

$$P_{max} = P_{mech} + P_{loss} \tag{5.15}$$

$$P_{max} = 2\pi n T_{max} + R_A I_{A\,max}^2 \tag{5.16}$$

In the above equation (5.15), P_{loss} is the electrical power loss which is converted into heat and is independent of the rotational speed. Mechanical losses such as those caused by bearing and air-gap friction are ignored. The maximum power available at the output shaft P_{mech} and the maximum electrical input power into the circuit increase in direct proportion to the increasing rotational speed. The limit to the primary regulation range $V_{A\,max}$ is set by the voltage strength of the winding insulation.

Field control. Field control of speed is used when higher speeds are required than those available with armature control. However, in this case an increase in speed regulation and in the available load torque is experienced as indicated in the following equations, where the conditions set out are valid for field control:

$$V_A = V_{A\,max} = \text{constant} \tag{5.17}$$

$$\phi_{max} \geq \phi \geq \phi_{min} \tag{5.18}$$

From equation (5.1) we have the following relationship between rotational speed and magnetic flux ϕ:

$$n = \frac{1}{k_1 \phi}(V_{A\,max} - R_A I_A) \propto \frac{1}{\phi} \tag{5.19}$$

If the load torque is considered at the same time then from equation (5.7) we get:

$$n = \frac{V_{A\,max}}{k_1 \phi} - \frac{R_A}{k_1 k_2 \phi^2} T \tag{5.20}$$

In contrast to armature control where the slop of the graph $-R_A/(k_1 k_2 \phi^2_{max})$ is a constant, in field control there is no similar family of characteristics $T(n)$. As the flux decreases, the curves become steeper—$\propto 1/\phi^2$, i.e. the motor becomes 'softer', as may be seen in Fig. 5.5.

The maximum permissible continuous-load torque reduces as the rotational speed increases due to the reduced flux:

$$T_{max} = k_2 \phi I_{A\,max} \propto \phi \tag{5.21}$$

From equation (5.14) the power input is given by:

$$P_{max} = V_{A\,max} I_{A\,max} = k_1 \phi I_{A\,max} n + R_A I^2_{A\,max} \tag{5.22}$$

$$P_{max} = \text{constant} \tag{5.23}$$

$$P_{loss} = \text{constant} \tag{5.24}$$

and with $n \propto 1/\phi$:

$$P_{mech} = k_1 \phi n I_{A\,max} = \text{constant} \tag{5.25}$$

A characteristic feature of field control is that the maximum continuous output power at the shaft P_{mech} and the maximum electrical power input into the system are constant (given that the minimal energizing losses are ignored).

In order to improve the speed regulation of main drives, particularly for field control, motors are today often fitted with a speed-control unit (see Volume 3). The actual speed is determined by a tachogenerator. The stator voltage and/or the exciting current is then altered until the deviation error equals zero, i.e. the selected set speed equals the actual speed (see Fig. 5.7).

Figure 5.6 shows a picture of and a sectional drawing through a DC machine which is self-cooling. The manufacturer of this design, which incor-

213

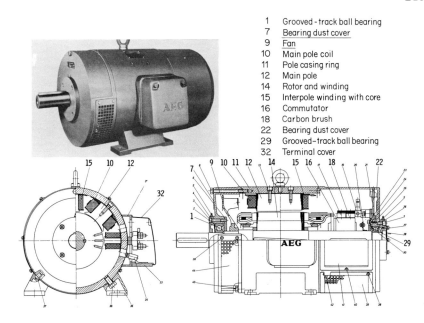

1	Grooved-track ball bearing
7	Bearing dust cover
9	Fan
10	Main pole coil
11	Pole casing ring
12	Main pole
14	Rotor and winding
15	Interpole winding with core
16	Commutator
18	Carbon brush
22	Bearing dust cover
29	Grooved-track ball bearing
32	Terminal cover

Fig. 5.6   Photograph and sectional drawing of a DC machine (AEG)

porates a built-in fan, offers machines with nominal speeds ranging from 500 to 3000 rev min^{-1}. The nominal power range is from 0.5 to 160 kW and the nominal torque from 3 to 1000 N m. If it is desired to have a control fitted for constant speed, a tachogenerator may be fitted on the brush end of the shaft. In order to minimize heat losses arising from eddy currents, the rotor and stator winding cores are made from insulated laminations.

Apart from separately excited DC machines, where the rotor and excited coils have separate supply currents, other circuits (e.g. shunt-wound and series-wound motors) are also in use, but these are of little importance in machine-tool construction.[66]

Current rectification for provision of direct current. The energy required to drive DC machines is today mostly obtained from alternating current, single or three phase, by using semi-conductor rectifiers. Either uncontrolled rectifiers (diodes) or controlled rectifiers (thyristors, triacs, transistors) may be employed. Figure 5.7 is a block circuit diagram of a DC drive fitted with speed control and a secondary current-control circuit. The setting of the rotor current for the separately excited motor is obtained by means of a 'phase gate' for 'intersection' control, the principle of which is shown in the upper left of Fig. 5.7. Depending on the triggering signal on the gate electrodes of the thyristors, a differing part of the voltage half-wave of the alternating circuit is switched in, so that in conjunction with the smoothing chokes a variable DC voltage is made available. The field windings are fed from a separate rectifier.

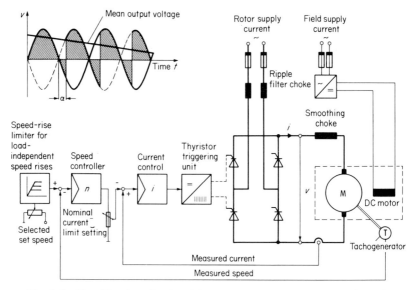

Fig. 5.7 Rectification circuit with controlled current rectifiers (AEG)

The rotational speed is monitored by the tachogenerator and compared with the required value. The deviation is transmitted through the speed controller in the form of a current-limiting value to the current controller. The triggering of the thyristors is then modified in accordance with the current deviation. Due to the provision of the secondary current-control circuit, it is possible to compensate for varying load torques before these cause changes in rotational speed. In the event of an increase in speed in the 'no-load' condition, the pre-set maximum nominal-speed value may be fed into the speed controller by a speed-limitation unit.

A more complex circuit is used to operate rectifier units from three-phase alternating current mains.

5.1.1.2 *Synchronous machines*

Construction and working principles. The stator of a synchronous machine has one or more windings carrying single- or three-phase alternating current which are distributed symmetrically around its circumference whilst the rotor is supplied with a DC voltage via slip rings. The rotor current, in contrast to the DC machine, is maintained in one direction so that the rotor is provided with fixed north and south poles, just as in a permanent magnet. Figure 5.8 is a diagram of the principle of operation. As the magnetic flux in the stator changes in direction due to inversions of the AC supply, so the rotor is caused to make one revolution in each cycle time t due to the changing attraction and repulsion effects, i.e. it rotates in synchronization with the frequency of the

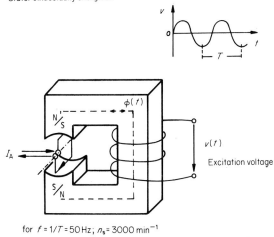

Fig. 5.8 Principle of operation for a synchronous machine

supply current (synchronous speed n_s). When:

$f = 50$ Hz, then $n_s = 50$ s$^{-1} \equiv 3000$ rev min^{-1}

If several separate windings are distributed around the circumference, the synchronous speed reduces (each winding is one pair of poles p):

$$n_s = \frac{f}{p} \tag{5.26}$$

As the machine can only develop a unidirectional torque at a particular rotational speed in accordance with the number of pairs of poles and the frequency of the supply voltage, it is not self-starting and must have a starting device (e.g. a starting motor or an asynchronous winding). The absolute consistency of the rotational speed is a particular feature of these motors. If a torque overload occurs, $(T > T_s)$ the motor goes 'out of step' and stops.

Variation of rotational speed. If it is necessary for a synchronous machine to operate at differing rotational speeds, this may be arranged by varying the frequency of the supply current. In the upper part of Fig. 5.9 a static frequency changer for an AC supply is diagrammatically presented. The 50 Hz supply voltage (phases R, Y, B) is first rectified. From the direct current, the subsequent DC/AC converter generates an three-phase system R', Y', B', the frequency of which, f', is variable.

If the three stator windings in a three-phase synchronous motor are arranged in slots, as shown in the lower part of Fig. 5.9, then in accordance

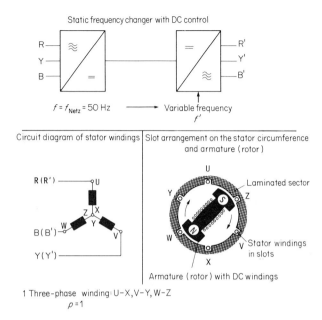

Fig. 5.9 Speed control for a synchronous machine

with equation (5.26) the following relationship exists, governed by the number of pole pairs of the motor:

$$n'_s = \frac{f'}{p} \quad \text{when driven by R', Y', B'}$$

The main application of synchronous machines is for the production of electrical energy in generating stations. For example, when driven by water or steam turbines, synchronous generators running at constant rotational speed produce alternating current for the public electricity-supply network. Owing to the disadvantages mentioned above, synchronous machines, as motors, are only employed for special duties, e.g. in such instances where the rotational speed must be accurately maintained, irrespective of the loading condition (gear grinding with grinding worms, see Volume 1).

5.1.1.3 Asynchronous (induction) machines

Construction and working principles. The stator of an asynchronous machine has a similar construction to that of a synchronous machine, but the rotor is not connected to an external supply source. The rotating field of the stator induces a voltage into the conductors of the rotor. This produces a current whose magnetic field, together with the stator field, produces a torque on the rotor. The smaller the difference (slip) between the rotor speed and the speed of the stator field, the smaller will be the induced current. The body of the armature consists of insulated iron laminations in order to minimize eddy currents and losses due to magnetic reversals.

The relationship between rotational speed and torque of an asynchronous machine is shown in Fig. 5.10. The most suitable operating range of such a motor is just below the synchronous speed, where the speed remains nearly constant under load, this being a basic requirement for a main drive motor. The starting torque is low, and until the 'pull-out' speed is reached the motor torque increases with the speed. The synchronous speed n_s which corresponds to the speed of the rotating stator field is only reached under 'no-load' conditions. As the load increases, so the slip increases:

$$s = \frac{n_s - n}{n_s} \qquad (5.27)$$

i.e. the difference between the synchronous speed and the actual speed. Under normal running conditions, the slip $s_N = 3-5\%$. Apart from the low starting torque, a further disadvantage arises from the high starting current, which may reach up to five times the nominal current.

One method available to reduce the current when starting is to utilize the 'star-delta' switching system of the stator windings shown in Fig. 5.11. When the star connection is in use during the starting period, the stator windings are supplied with 220 V. Consequently, the current rises to only $1/\sqrt{3}$ of the value which would apply if a delta connection were used. However, the starting torque is reduced by a similar ratio, and therefore the star connection can only be applied under light loading, without any significant starting or friction torques.

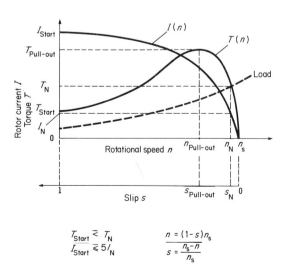

Fig. 5.10 Torque and rotor current curves of an asynchronous machine

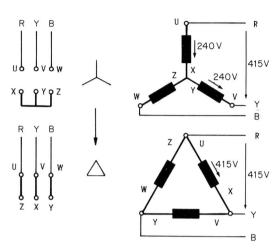

Fig. 5.11 Star-delta switching of stator windings

Types of rotors. The behaviour of an asynchronous machine is largely governed by the type of rotor winding. On a slip-ring rotor a three-phase winding is wound on to the laminated rotor body and the ends are connected to the slip rings and brushes. By connecting controlling resistors as shown in Fig. 5.12, the rotor current may be limited. The starting torque is increased and the pull-out torque is reached at a lower rotational speed. The torsional stiffness in the operating range reduces as the value of the controlling resistors

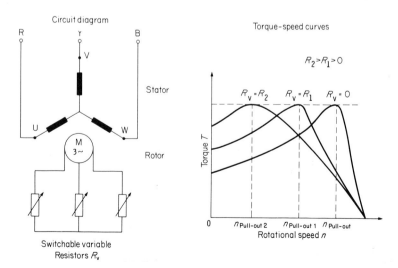

Fig. 5.12 Slip-ring rotor motor

is increased. In order to be able to reach the required shunt characteristics for the rating of the motor, the controlling resistors are short-circuited when the maximum speed has been attained ($R_V = 0$).

In the case of the squirrel-cage motor, also known as a short-circuit rotor motor, the winding is replaced by copper or aluminium rods fitted into the slots of the rotor and short-circuited at both ends with conducting rings. The body of the rotor is again made from insulated iron laminations. As a result of the simple construction and the absence of components prone to wear, such as brushes and slip rings, these motors are known for their robustness and comparatively low cost.

Figure 5.13 shows a three-phase standard motor with a squirrel-cage rotor. Depending upon the construction of the stator windings, this type of motor can have nominal speeds (n_N) from 480 rev min^{-1} ($n_s = 500$ rev min^{-1}) to 2800 rev min^{-1} ($n_s = 3000$ rev min^{-1}) when operating from a 50 Hz supply frequency. Nominal power ratings may range from 0.5 to 300 kW according to motor size. To improve the conduction of heat losses, the motors, which are also frequently available as flange motors, are provided with ventilator fans and ribbed casings. Fixing-hole centres and axis heights of the various size ranges are governed by DIN, BSI and IEC standard specifications.

The operating characteristics of a squirrel-cage motor are governed by the cross-section of its rotor conductors. Figure 5.14 shows widely used cross-section shapes and the corresponding torque–speed curves. The various characteristics are a result of the current displacement in the conductors. During the starting period, i.e. when there is considerable slip, the frequency of the induced rotor voltage is high. At these frequencies, only the outer part of the cross-sectional area of the conductors conducts current. Consequently, the conductor bars have a high resistance resulting in a reduced starter current and an increased torque (giving an effect similar to that of the control

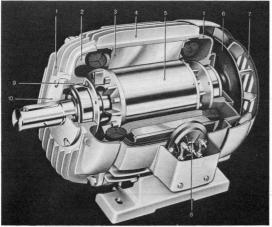

1 Bearing cover
2 Stator winding
3 Stator lamination stack
4 Casing
5 Armature
6 Fan
7 Fan cover
8 Terminal plate
9 Inner bearing plate
10 Outer bearing plate

Fig. 5.13 Three-phase standard motor with squirrel-cage rotor (AEG)

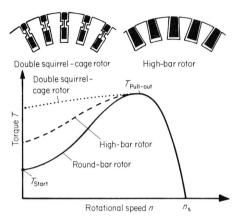

Fig. 5.14 Rotating-speed–torque relationships for different types of armature

resistors in the slip-ring rotor motor). As the rotational speed increases, and hence the rotor-current frequency decreases, current flows through the whole of the conductor cross-section so that a good speed regulation can be guaranteed in the nominal working-speed range.

In order to obtain a variable range of working speeds an asynchronous motor is usually fitted with a subsequent gear drive or stepless transmission unit. A continuously variable motor speed is only possible with the use of a frequency changer unit, as in the case of synchronous machines (see section 5.11.2 and Fig. 5.9).

The continuously reducing costs in the field of electronic products is, in the near future, expected to lead to an ever-increasing importance of squirrel-cage motors for the main drives of machine tools.

5.1.2 *Hydraulic motors*

Apart from electric motors, hydraulic motors are also used in machine tools for the main and feed motions. The considerably increased power density which is a feature of hydraulic motors is the main advantage over their electrical counterparts. This enables smaller overall construction sizes to be obtained with consequential weight reductions and quicker manufacture. Against these are the disadvantages of hydraulic drives, which include a low efficiency, the sensitivity to dirt, the effect of temperature variations and, as a result of the high power density coupled with friction, the rapid wear of some individual constructional elements.[67–70]

Figure 5.15 gives a diagrammatic presentation of the components of a hydraulic installation. Electrically driven oil pressure pumps establish an oil flow for energy transmission, which is fed to hydraulic motors or hydraulic cylinders, converting it into mechanical energy. The control of the oil flow is

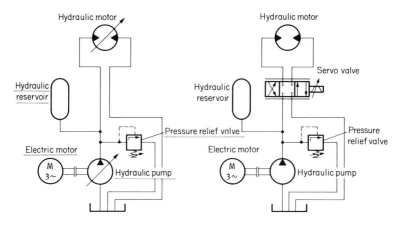

Fig. 5.15   Speed-control methods for hydraulic motors

Three alternative methods of control
1. V_P 2. V_M 3. Servo valve

by means of valves. The pressurized oil flow produces linear or rotary mechanical motion. The kinetic energy of the oil is comparatively low, and therefore the term 'hydrostatic drives' is sometimes used. There is little constructional difference between hydraulic motors and pumps. Any pump may be used as a motor. The quantity of oil flowing at any given time may be varied by means of regulating valves (Fig. 5.15) or the use of variable-delivery pumps.

In general terms, hydraulic drives may be divided into rotary and linear types. Rotary drives produce a rotating motion, whilst linear devices in the form of piston and cylinder units produce a reciprocating movement. The construction and operating principles of a number of types of motor are described in the following sections.

All hydraulic motors function broadly in accordance with the same basic principle. A pressurized fluid is alternately forced into and removed from a chamber. The filling cycle begins with minimum chamber volumes. When the chamber reaches its maximum volume (the maximum capacity) the filling cycle is ended by isolating the chamber from the supply lines. The oil is then returned to the oil sump through the return lines and at the same time the next chamber is filled with oil.

The basic relationship necessary to understand the functioning of a hydraulic motor are presented, based upon a simplified model of a displacement machine, shown in Fig. 5.16. The swept volume per revolution of the design shown in the illustration is:

$$V = A_{piston} h = \frac{d^2 \pi}{4} h \qquad (5.28)$$

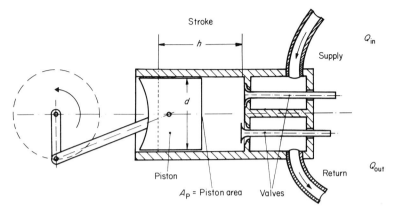

Fig. 5.16 Model of a hydraulic displacement machine

From this we obtain the oil flow per unit time as:

$$Q = nV \tag{5.29}$$

As the same conditions apply to the pump as well as the motor and as the motor receives the whole of the flow transmitted from the pump ($Q_P = Q_{Mot}$) then, assuming there are no losses, the speed relationship of a hydraulic drive is given by:

$$\frac{n_{Mot}}{n_P} = \frac{V_P}{V_{Mot}} \tag{5.30}$$

and the motor speed is:

$$n_{Mot} = \frac{V_P}{V_{Mot}} n_P \tag{5.31}$$

The motor power output is given by:

$$P_{Mot} = V_{Mot} n_{Mot} p_{mean} = \frac{d^2 \pi}{4} h n_{Mot} p_{mean} \tag{5.32}$$

or from:

$$Q = n_{Mot} V_{Mot} \tag{5.33}$$

$$P_{Mot} = Q p_{mean} \tag{5.34}$$

From this equation the mean torque may be obtained. In general, the following is valid:

$$P = T\omega \tag{5.35}$$

where:

$$\omega = 2\pi n \tag{5.36}$$

The motor power output is then:

$$P_{Mot} = T_{mean} 2\pi n_{Mot} = V_{Mot} n_{Mot} p_{mean} \qquad (5.37)$$

From this we obtain the mean torque as:

$$T_{mean} = \frac{1}{2\pi} V_{Mot} p_{mean} \qquad (5.38)$$

As this equation is valid for both the pump and the motor, the torque transmission between motor and pump when $p_P = p_{Mot}$ is given by:

$$\frac{M_{Mot}}{T_P} = \frac{V_{Mot}}{V_P} \qquad (5.39)$$

and the motor torque by:

$$T_{Mot} = T_P \frac{V_{Mot}}{V_P} \qquad (5.40)$$

5.1.2.1 Rotary hydraulic drives

Gear motors. In a gear motor as shown on the left of Fig. 5.17 the tooth spaces combine with the casing to form the displacement spaces. The separation between pressurized and non-pressurized spaces is obtained by the meshing of the teeth and between their crests and the inner casing wall. It is not possible to vary the volume of gear motors. The capacity volume per revolution is given by:

$$V = 2\pi z b m^2 k \qquad (5.41)$$

where m = module
 z = number of teeth
 b = tooth width
 k = a correction factor to allow for the tooth spaces being larger than the teeth

The maximum pressures available are between 160 and 200 bar.

On the right of Fig. 5.17 an internal-gear motor is illustrated. The pinion rotor with the external teeth has one tooth less than the internal-gear rotor. Both rotors revolve about parallel axes at fixed centres. The theoretical displacement volume per revolution is given by:

$$V_{th} = Ab \qquad (5.42)$$

where A = the free cross-sectional area (measured with a planimeter) between the internal-gear and the pinion
 b = tooth width

Pressures up to $p = 250$ bar are attainable. To separate the pressurized and de-pressurized sides a control plate is used.

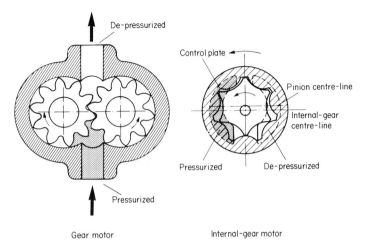

Fig. 5.17 Gear motor and internal-gear pump

Vane-type motors. In this type of motor the displacement volume is obtained from vanes which slide in radial slots in the rotor, in conjunction with the rotor itself and the casing. In Fig. 5.18 an externally fed motor and an internally fed motor are shown on the left and right respectively. This classification refers to the method of supply of the pressurized oil. In the former case this is arranged externally and in the latter the fluid is fed through the centre of the rotor shaft. The variation in speed and maximum available torque is obtained by adjustment of the eccentricity e. Pressures up to $p = 100$ bar are attainable with circular stator designs, as shown in Fig. 5.18. The maximum displacement volume is obtained when the eccentricity is at a maximum, i.e.:

$$e_{max} = \frac{D - d}{2} \qquad (5.43)$$

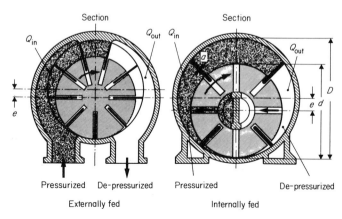

Fig. 5.18 Vane-type motor

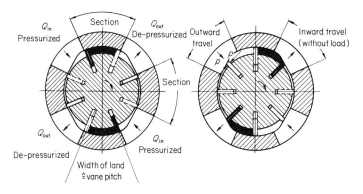

Fig. 5.19 Vane-type motor (Teves)

and is then

$$V_{max} = b\left[\frac{\pi(D^2 - d^2)}{4} - \frac{az(D - d)}{2}\right] \quad (5.44)$$

where b = width of vane
        a = thickness of vane
        z = number of vanes

The displacement volume may be varied by changing the eccentricity e and is given by:

$$V = V_{max}\frac{e}{e_{max}} \quad (5.45)$$

A different design of vane-type motor is shown in Fig. 5.19. By a symmetrical arrangement of the housing, the rotor is relieved of all radial stresses. As the rotor is not adjustable relative to the stator, no speed variation is possible within the motor itself. Another special feature of this motor design is that the vanes will only move linearly when they are de-pressurized (i.e. the same pressure is acting on both sides of the vane). The stator has a cam profile only in the sections of oil inlet and outlet. Hence there is only low friction in the guides, so that peak pressures of up to $p = 210$ bar are obtainable.

The widths of the lands must be at least equal to one vane pitch in order to prevent a back flow between the pressurized and non-pressurized sides.

Piston-displacement units. The displacement volumes in these motors are produced by pistons and cylinders, the various arrangements of which result in a range of design types.

Axial piston motors. In the case of axial piston motors the drive shaft and the piston axis are parallel to each other. Figure 5.20 shows an axial piston motor with swash plate. The piston drum and its pistons rotate, and due to contact with the swash plate an axial oscillation of the pistons is activated. The

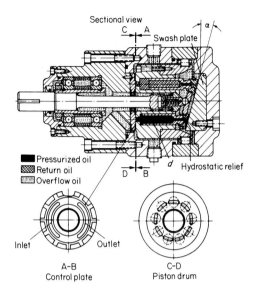

Fig. 5.20 Axial piston pump (Denison)

displacement volume per revolution of the motor is given by:

$$V = \frac{d^2\pi}{4} z(2r \tan \alpha) = \tfrac{1}{2}\pi d^2 rz \tan \alpha \tag{5.46}$$

where d = piston diameter
 z = number of pistons
 r = radius of contact circle of pistons on swash plate
 α = angle of swash plate
 $(2r \tan \alpha)$ = piston stroke

The angle of the swash plate shown in Fig. 5.20 is not variable. The speed of variable speed motors is altered by changing the angle of the swash plate. As the pump provides a constant delivery, Q = constant, we have a relationship:

$$n_{\text{Mot}} \propto \frac{1}{V_{\text{Mot}}} \tag{5.47}$$

and

$$T_{\text{Mot}} \propto V_{\text{Mot}} \tag{5.48}$$

where $V_{\text{Mot}} = f(x)$ and $\alpha \neq 0$ as otherwise $n_{\text{Mot}} \to \infty$.

Figure 5.21 shows an axial piston motor with an adjustable swash plate. The setting of the angle is carried out by means of a hydraulic slave cylinder which provides the large forces necessary to move the swash plate with relatively low applied pressures.

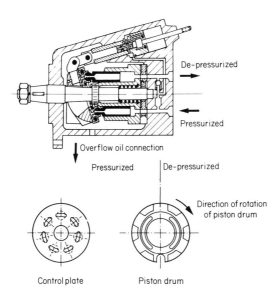

Fig. 5.21 Axial piston motor with adjustable swash plate (Bosch)

Radial piston motors. In the case of radial piston motors the piston axis and the axis of the drive shaft are perpendicular to each other. When the motor has the fluid supplied and returned from ports at its centre (internally fed) the main components are a casing and a rotor. These rotate about parallel axes, the relative centres of which are adjustable. The housing rotates in the same direction as the rotor, which tends to reduce the frictional effects between the pistons and the casing. The alternative design of motor, having the fluid supplied and returned from external ports (externally fed), has the pistons supported by an internal shaft which is fitted with a rolling bearing to reduce friction. Speed variations and changes in direction of rotation are achieved by variations in the eccentricity e or adjustments to the delivery of the pump:

$$V = \frac{\pi}{2} d^2 z e \rightarrow V = f(e) \tag{5.49}$$

where d = diameter of pistons
z = number of pistons
e = eccentricity

The function of this motor is clarified in Fig. 5.22. The motors are internally or externally fed in accordance with their basic design. A control plate is necessary to separate the fluid supply and return sides in the case of externally fed motors.

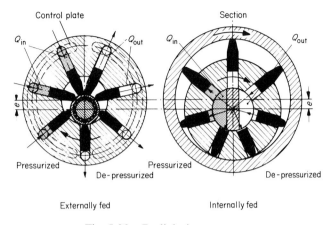

Fig. 5.22 Radial piston motors

5.1.2.2 Linear hydraulic drives

Another widely used hydraulic motor type is the linear drive. The main application is for hydraulic presses, slotting machines and shaping and planing machines where a piston and cylinder drive is used without the need for a mechanical transmission. Figure 5.23 shows a hydraulic table drive. A pump driven by a three-phase electric motor delivers the pressurized oil, the quantity of which is controlled by a valve dependent upon the table velocity which is required at any given time. The use of such a drive produces a constant cutting velocity for almost the whole length of the stroke. As a safety measure against overload, a pressure relief valve is fitted. The pilot valve control shown on the left in Fig. 5.23 (the main control valve is activated by a small pilot valve) reduces the necessary reversing forces, and the resultant reversal is softer (less shock). However, the throttling of the supply flow results in large power losses.

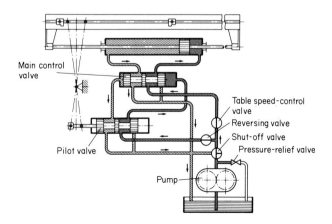

Fig. 5.23 Hydraulic table drive with pilot valve

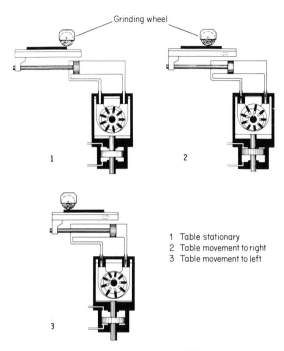

Fig. 5.24 Hydraulic table drive (Waldrich Coburg)

Another example of a fully hydraulic table drive is shown in Fig. 5.24. The housing of the vane pump is moved by means of a piston, and this enables the flow quantity and the direction of flow to be changed. An advantage of linear hydraulic drives is that there are no fast rotating masses which must be retarded and accelerated at each table reversal.

5.1.2.3 Speed control
The speed control of hydraulic motors may be achieved by one of three methods:

1. Variation of the delivery volume by altering the pump volume (primary variations); Fig. 5.25.
2. Variation of the swept volume of the motor (secondary variation); Fig. 5.26.
3. Variation of the delivery volume by a control valve (normally an electrically operated valve); Fig. 5.27.

Each of the above methods follow different relationships, as detailed below.

Characteristic relationships for method 1 (Fig. 5.25) The variation of the delivery volume of the pump is produced by adjusting the eccentricity e in the case of vane and radial piston pumps and by adjusting the angle of inclination

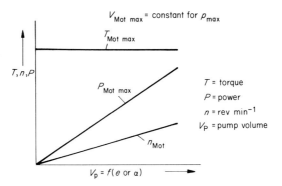

Fig. 5.25 Characteristic curves of a hydraulic motor with pump volume adjustment

α of the swash plate in the case of axial piston pumps:

(a) $V_{\text{Mot max}} = $ constant (5.50)

(b) $V_P = f(e \text{ or } \alpha)$ (5.51)

(c) $Q_P = V_P(e \text{ or } \alpha) n_P$ (5.52)

(d) $Q_P = Q_{\text{Mot}}$ (5.53)

$\qquad V_P(e \text{ or } \alpha) n_P = V_{\text{Mot}} n_{\text{Mot}}$ (5.54)

$\qquad \to n_{\text{Mot}} = n_P \dfrac{V_P(e \text{ or } \alpha)}{V_{\text{Mot}}}$ (5.55)

(e) $P_{\text{Mot max}} = V_P(e \text{ or } \alpha) n_P p_{P\,\text{max}}$ (5.56)

(f) $T_{\text{Mot max}} = \dfrac{V_{\text{Mot}} p_{P\,\text{max}}}{2\pi} = $ constant (5.57)

Characteristic relationships for method 2 (Fig. 5.26). The swept volume of the motor is varied by adjusting the eccentricity e in the case of vane and radial piston motors and by adjusting the angle of inclination α of the swash plate in the case of axial piston motors. When $V_P = $ constant and $Q_P = $ constant, i.e. the pump remains unaltered, the following relationships are varied:

(a) $V_{\text{Mot}} = f(e \text{ or } \alpha)$ (5.58)

(b) $b_{\text{Mot}} = \dfrac{Q}{V_{\text{Mot}}(e \text{ or } \alpha)}$ (5.59)

(c) $P_{\text{Mot max}} = Q p_{P\,\text{max}} = T_{\text{Mot max}} 2\pi n_{\text{Mot}} = $ constant (5.60)

(d) $T_{\text{Mot max}} = \dfrac{P_{\text{Mot max}}}{2\pi n_{\text{Mot}}} = \dfrac{1}{2\pi} p_{P\,\text{max}} V_{\text{Mot}}(e \text{ or } \alpha)$ (5.61)

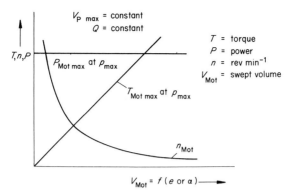

Fig. 5.26   Characteristic curves of a hydraulic motor
with motor swept-volume adjustment

where (c) and (d) are limiting-load relationships. By analogy to a separately excited DC motor:

For primary variations:

$V_A \mathrel{\hat{=}} V_P(e \text{ or } \alpha)$
when V_{Mot} = constant
(see section 5.2)

For secondary variations:

$\phi_{Mot} \mathrel{\hat{=}} V_{Mot}(e \text{ or } \alpha)$
when Q = constant

Characteristic relationships for method 3 (Fig. 5.27). Due to the flow characteristics of the control valve there is no linear relationship between speed and

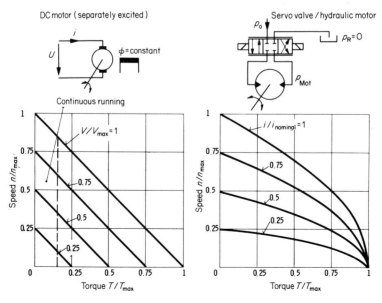

Fig. 5.27   Comparison between DC motor and hydraulic motor

torque in a valve-controlled constant-volume hydraulic motor. A diaphragm effect is experienced on the control valve edges:

$$Q \propto i\sqrt{(p_{\text{Val}})} = i\sqrt{(p_P - p_{\text{Mot}})} \tag{5.62}$$

where

$$p_{\text{Mot}} = \frac{2\pi}{V_{\text{Mot}}} T_{\text{Mot}} \tag{5.63}$$

The control valve flow i produces a proportional movement of the control piston (increase in cross-sectional area). If the swept volume of the motor is constant, then:

$$n = \frac{Q}{V_{\text{Mot}}} \tag{5.64}$$

$$n = Ci\sqrt{\left(p_P - \frac{2\pi}{V_{\text{Mot}}} T\right)} \tag{5.65}$$

In the above, C is a motor/valve constant.
Since

$$n_{\max} = Ci_N\sqrt{p_P}$$

and

$$p_P = \frac{2\pi T_{\max}}{V_{\text{Mot}}}$$

we have, when considered as a specific ratio in relation to maximum speed and nominal flow:

$$\frac{n}{n_{\max}} = \frac{i}{i_N}\sqrt{\left(1 - \frac{T}{T_{\max}}\right)} \quad \text{if} \quad p_P = \text{constant} \tag{5.66}$$

This relationship is depicted on the graph shown on the right of Fig. 5.27. The corresponding relationship of a separately excited DC motor is shown for comparison on the graph reproduced on the left of the diagram. A linear relationship between speed and torque exists for separately excited DC motors:

$$n = \frac{1}{k_1 \phi} V - \frac{R_A}{k_1 k_2 \phi^2} T \tag{5.67}$$

When considered as a specific ratio in relation to maximum speed, voltage and torque, we have:

$$\frac{n}{n_{\max}} = \frac{V}{V_{\max}} - \frac{T}{T_{\max}} \tag{5.68}$$

The time constant when starting is independent of the armature voltage as well as the final speed:

$$C_t = \frac{2\pi n_{max} J_{tot}}{T_{max}} \tag{5.69}$$

(see section 5.1.3). As the armature mass is large in relation to the torque, the time constant t is usually greater than in a comparable hydraulic drive.

5.1.3 Start-up conditions of a drive

So far we have considered the static characteristics of motors which are valid for given settings after the starting period and under static operating conditions. We now examine the start-up or acceleration period of a drive from the moment of switching on.

In the example a separately excited DC motor having a characterisic speed–torque curve T_M (Fig. 5.28) drives a load T_L. The total relative moment of inertia of all moving masses is designated J_{tot}. According to the equation of motion for rotation we have:

$$T_B(\omega) = T_M(\omega) - T_L(\omega) = J_{tot} \dot{\omega} \tag{5.70}$$

where:

$$\omega = 2\pi n \tag{5.71}$$

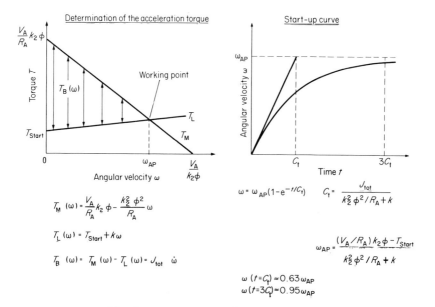

Fig. 5.28 Start-up conditions of a drive

T_B is the torque which is required for the motor to accelerate the moving masses. Its magnitude is derived from equations (5.1) and (5.2) which produce the following function in terms of the angular velocity ω:

$$T_B(\omega) = \left(\frac{V_A}{R_A}k_2\phi - \frac{k_2^2\phi^2}{R_A}\omega\right) - (T_{Start} + k\omega) = J_{tot}\dot{\omega} \quad (5.72)$$

After rearrangement a linear differential equation of the first order is obtained:

$$C_t\dot{\omega} + \omega = \omega_{WP} \quad (5.73)$$

$$\frac{J_{tot}}{k_2^2\phi^2/R_A + k}\dot{\omega} + \omega = \frac{(V_A/R_A)k_2\phi - T_{St}}{k_2^2\phi^2/R_A + k} \quad (5.74)$$

The solution to this differential equation is:[71]

$$\omega(t) = \omega_{WP}(1 - e^{-t_1/C_t}) \quad (5.75)$$

By comparing the coefficients in equations (5.73) and (5.74) we obtain the time constant C_t for the working speed to be reached and the angular velocity ω_{WP} which is reached at the working point:

$$C_t = \frac{J_{tot}}{k_2^2\phi^2/R_A + k} \quad (5.76)$$

$$\omega_{WP} = \frac{(V_A/R_A)k_2\phi - T_{Start}}{k_2^2\phi^2/R_A + k} \quad (5.77)$$

The start-up curve follows the form shown on the right of Fig. 5.28 (e function).

If the curve for motor speed and torque is non-linear then an approximation may be used to determine the start-up relationships (as shown in Fig. 5.29). Within a given angular velocity range $\Delta\omega_i$, the actual values of $T_M(\omega)$ and $T_L(\omega)$ are replaced by their mean values $\bar{T}_M(\Delta\omega_i)$ and $\bar{T}_L(\Delta\omega_i)$ (stepped curve). The equation of motion thus becomes linear for these velocity ranges:

$$\bar{T}_B(\Delta\omega_i) = \bar{T}_M(\Delta\omega_i) - \bar{T}_L(\Delta\omega_i) = J_{tot}\frac{\Delta\omega_i}{\Delta t_i} \quad (5.78)$$

In this way the time required Δt_i for the motor to pass through the angular velocity range $\Delta\omega_i$ can be calculated:

$$\Delta t_i = \frac{J_{tot}\Delta\omega_i}{\bar{T}_B(\Delta\omega_i)} \quad (5.79)$$

By adding up the time values Δt_i on the time axis, the start-up characteristic curve can be constructed. Providing the intervals $\Delta\omega_i$ are sufficiently small, the accuracy of this method is quite adequate.

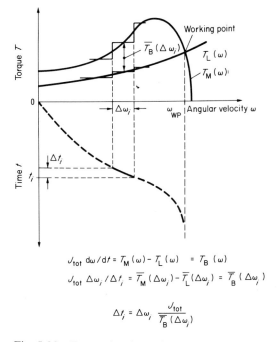

Fig. 5.29 Determination of start-up relationships $\omega(t)$ for non-linear speed–torque curve

5.2 Transmission drives

5.2.1 General requirements

A transmission drive is basically a device which accommodates and positively converts movements, torques and forces during energy transfer. In machine-tool construction[67–69] transmission drives are mainly used to convert the generally high rotational speeds of the motors into useful cutting velocities and furthermore to produce the necessary feed motions of the tool holders.

Transmission drives may be classified into uniform and non-uniform types (Fig. 5.30). In a uniform transmission drive the input and output motions are in a fixed relationship to each other. The uniform transmission drives may be further subdivided into stepped and stepless units. This means that drives either produce a finite number (steps) of intermediate speeds between their minimum and maximum speed (stepped drives) or, within that range, any intermediate speed may be set continuously (stepless drive). Stepless drives enable, for example, the maintenance of optimum machining speeds for different production conditions on cutting machine tools. However, their lower efficiency and greater bulkiness when compared with stepped drives of similar power capacity must be considered as disadvantages.

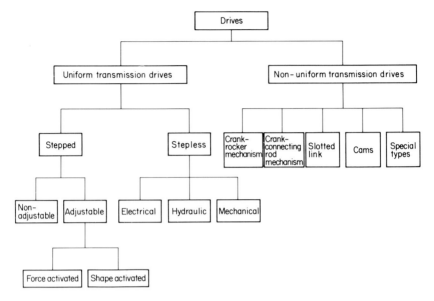

Fig. 5.30   Classification of drives

The non-uniform transmission drives produce a varying working velocity from a constant angular velocity input. In general, the drive output is linear and the velocity v is given by $\delta v/\delta t = \dot{v} \neq 0$. The variation of speed may be obtained in either a stepped or stepless form.

5.2.2   Uniform transmission drives

5.2.2.1   Drives with stepped-speed changes

The stepped transmission drives include both those with fixed (not changeable) and variable (changeable) transmission ratios. In machine-tool construction stepped transmission drives are used in the form of gear boxes or belt drives (two-speed or variable-speed motors do not come into the category of transmission drives). The simplest form of a gear drive consists of a single pair of gears, where the driving and driven gear are fitted to different shafts. The two meshing gears usually have different effective diameters, so that the rotational speeds of the shafts have a definite ratio. A variable transmission drive has various wheel pairings available for selection, and according to their arrangement a number of basic designs have been established.

Basic design types. A number of devices is available for operating the change mechanism in drives, e.g. sliding gears, mechanical or electrical clutches, etc. (see section 5.3). The simplest form of changeable gear drive is a basic two-step unit with two output speeds, and the next stage is a three-step unit

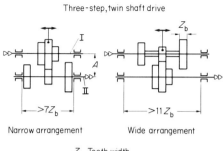

Fig. 5.31 Three-step basic gear drive units

providing three output speeds for selection, as shown in Fig. 5.31. The change is provided by sliding gears in the designs illustrated. The upper three gears run at a constant speed on their shaft, but may be moved along their axis of rotation. In one sliding gear drive, the gears are able to move on both the input as well as the output shaft. It is advantageous to move the smaller gears as these are of a smaller mass and will need shorter selector forks in the mechanism.

The narrow arrangement shown in the diagram is preferred as this leads to smaller bulk than the alternative wide arrangement. In the latter case, the axial length is longer and at the same time the shafts must be of larger diameters to withstand the higher bending and torque loadings. This means that the smallest wheel must be larger in diameter with a consequential increase in the size of all the other gears. In this arrangement the largest wheel is placed in the centre of the moving gear cluster. The difference between the largest wheel diameter and that of the second largest must be such as to permit the moving cluster to pass unhindered over the central fixed gear wheel. Consequently it is not possible to have small steps in output speeds in a three-step basic gear drive.

By arranging several basic drives in series, gear drives with larger numbers of output speeds may be obtained. Two two-stepped basic units are shown in series as an example in Fig. 5.32, resulting in a four-step gear drive with four output speeds.

To reduce the length of the gear box and minimize the gears required, idler gears may be employed. In this way, one or more gears are used in more than one gear train. The idler gears are shown shaded in Fig. 5.32. As the idler gears have to engage with two other wheels at the same time, all three must have the same module. The size of the module is determined by the gear train loaded with the highest torque, which can lead to larger shaft-centre distances in certain circumstances. Hence the reduction in axial length of the gear box leads to an increase in radial size.

A widely used form of construction is the back gearing arrangement shown in Fig. 5.33. Such a gear box consists of three shafts and is operated by a

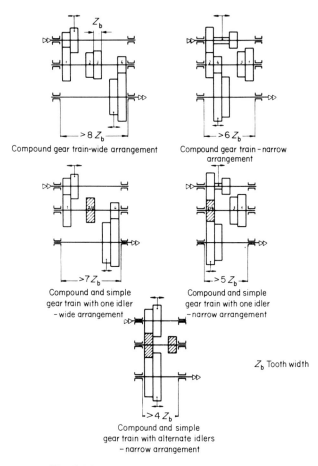

Fig. 5.32 Four-step drives on three shafts

coupling clutch. The power is transmitted either from shaft I directly to shaft III, or via shaft II on to shaft III.

In the former case, the gears remain in mesh but perform no function, and hence the input and output speeds are equal. In the latter case, because of the sequential engagement of two pairs of gears, a high final gear ratio is obtained. Due to this particular design (return of the power to shaft III in line with the input shaft) a reduced overall size of the gear box is obtained. Back gears are normally applied to input shafts so that inside the gear box high speeds and hence low torques are retained as long as possible (see later in this section).

Closed-loop drives such as that shown in Fig. 5.34 appear as two connected back gears with their shafts interchanged. However, after the speed change has taken place, the second train is not a back gear but a transmission drive.

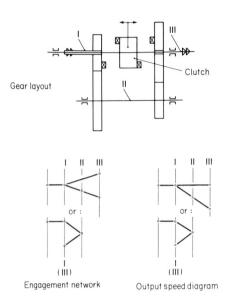

Fig. 5.33 Back gear arrangement (principles)

The input and output shafts are, in contrast to the back gear, not on the same axis.

Closed-loop gear drives are fundamentally two-axis units with several co-axial shafts. The gear boxes use the same number of gears as a drive using twin idler gears for comparable numbers of speed steps, but because the shafts are hollow they are more economical to build. The smallest gear on the hollow shaft can normally not have less than 20 teeth. Although this design is

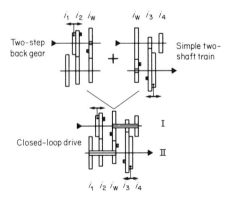

Fig. 5.34 Composition of a closed-loop drive from a back gear and a transmission gear train

comparatively long in the axial direction, a reduced space is required radially as only two shafts are employed.

Application examples. The transmission unit from a universal milling machine shown in Fig. 5.35 consists of two three-stepped basic gear units arranged in series, a belt drive and a back gear. The multi-vee belt drive on the output shaft reduces the shock loads and vibrations due to the interrupted cut, particularly at high speeds. The two three-stepped basic gear units provide nine speeds. With the introduction of the back gear a further six speeds may be obtained, so that a total of fifteen speeds are available for the main spindle. From the speed diagram (see Fig. 5.38) it may be seen that the speeds 140, 180 and 224 rev min^{-1} may be obtained directly from the belt drive, as well as with the back gear and the belt drive. The drive must be stationary when the sliding gears are engaged. The spindle is accelerated after a gear change by means of a multi-plate friction clutch (see section 5.33).

The main drive of a vertical milling machine, shown in Fig. 5.36, consists of a nine-speed clutch-operated gear unit, a belt drive and a two-stage final transmission drive. In contrast to sliding gears, clutch-operated gear units enable speed changes to take place while running under load. On the right of the flexible coupling between the motor and the main drive, a cooling unit may be seen which forces cool air into the gear box. Such devices are employed in heavy-duty machine tools to minimize the temperature rises induced by drive, clutch and bearing friction losses, thus transferring these energy

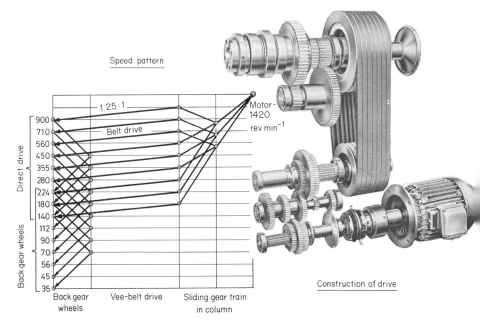

Fig. 5.35 Transmission unit and speed diagram of a milling machine (Heller)

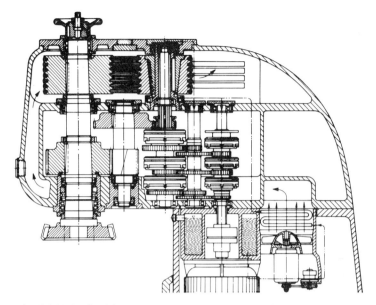

Fig. 5.36 Main-drive arrangement of a vertical milling machine (Heller)

losses into the cooling air. If this is not provided, unacceptably high thermal displacements may occur, particularly on the main spindle, which would impair the geometric accuracy of the machine (see section 2.5).

Figure 5.37 shows a design of a closed-loop drive, the principle of which is shown in Fig. 5.34. The gear wheels on the lower left and upper right are supported by bearings on the shafts, because gear clusters and shafts may have differing rotational speeds according to a particular gear selection.

Fundamentals for stepped-drive calculations. Standard machine tools are used for a wide range of production applications. The maximum speed ratio required is governed by the highest and lowest cutting velocity expected, as well as the largest and smallest diameters of the work or tool which will have to be accepted (see Fig. 5.38). The ratio between the highest and lowest working speeds is known as the 'maximum speed ratio' B. Apart from the maximum speed ratio, the number of speeds z (number of output speeds available for selection) is also of importance for the economic utilization of the machine.

For any given production condition an optimum speed may be determined at which the most economic cutting velocity will operate. In the case of stepped drives, the optimum speed can generally not be selected exactly, because of the fixed number of speeds available. The greater the number of speeds provided for selection, the closer will be any possible approximation. However, the smaller the intervals between speeds in a given speed range, the

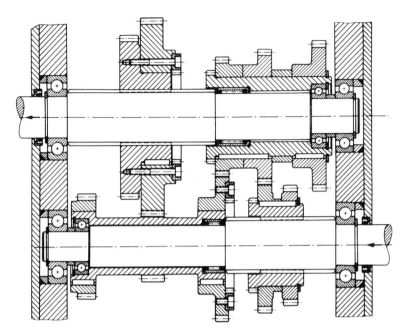

Fig. 5.37 Nine-speed closed-loop drive unit

greater will be the cost of the unit. Table 5.1 is a guide to the usual values of the maximum-speed ratio and the number of speeds ratio in machine tools.

Figure 5.39 presents the two principles which are applied to determine the speeds for a given speed range, i.e. the 'arithmetic progression' and the 'geometric progression'. Most main feed drives in machine tools follow a

$$B = \frac{n_{max}}{n_{min}}$$

$$n_{max} = \frac{v_{max}}{d_{min} \pi} \qquad n_{min} = \frac{v_{min}}{d_{max} \pi}$$

$$\frac{n_{max}}{n_{min}} = \frac{v_{max}}{v_{min}} \cdot \frac{d_{max}}{d_{min}}$$

$$B = B_v \cdot B_d$$

n_{max}	Highest speed	B	Maximum speed ratio
n_{min}	Lowest speed	B_v	Maximum velocity ratio
d_{max}	Largest work diameter	B_d	Maximum diameter ratio
d_{min}	Smallest work diameter		
v_{max}	Highest cutting velocity		
v_{min}	Lowest cutting velocity		

Fig. 5.38 Speed-change range

Table 5.1 Maximum speed ratios and number of speeds ratios

Type of machine	Maximum speed ratio	Number of speeds
Planing machines	6–10	6–9
Lathes	50–200	12–18
Milling and boring machines	Up to 400	Up to 36
Presses	1	1

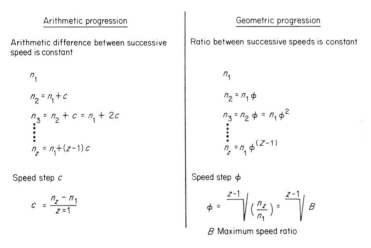

Fig. 5.39 Derivation of speed-step systems

geometric progression. An arithmetic progression is only practical for lead-screw drives intended for use in thread cutting where the pitches follow an arithmetic progression.

The feature of an arithmetic progression speed range is that the arithmetic difference between all successive speeds is constant. In a geometric progression, the ratio between all successive speeds is constant (constant-ratio geometric progression).

As mentioned earlier, the exact optimum speed is often not available and therefore the precise cutting velocity which would be most economical cannot be obtained. To evaluate a stepped-speed system, the concept of the 'cutting-velocity reduction ratio' shown in Fig. 5.40 is employed. This ratio, p, occurs when the speed is changed from one value to the next. For a given workpiece or tool diameter two cutting velocities v_u and v_l are applicable and from their difference the cutting-velocity reduction ratio p is calculated.

If the cutting velocity v is plotted against the cutting diameter d (work or tool diameter) as in Fig. 5.41, then for each speed n_1, n_2, n_3, etc., a linear relationship is developed with varying slope values ($v = \pi d n_i$). If an upper cutting velocity v_u is specified which corresponds to the ideal economic cutting velocity, then it may be seen that different speeds should be selected for

Cutting-velocity reduction ratio for a speed change

$$p_{(n)} = \frac{v_u - v_l}{v_u} \, 100\%$$

$v_u = d_2 \pi n_i$ $\qquad\qquad v_l = d_2 \pi n_{(i-1)}$

Arithmetic progression

$$p = \frac{n_i - n_{(i-1)}}{n_i} \, 100\% = \frac{n_i - (n_i - c)}{n_i} \, 100\% = \frac{c}{n_i} \, 100\% \longrightarrow p = f(c, n)$$

Geometric progression

$$p = \frac{n_i - n_{(i-1)}}{n_i} \, 100\% = \frac{n_{(i-1)}\phi - n_{(i-1)}}{n_{(i-1)}\phi} \, 100\% = \frac{\phi - 1}{\phi} \, 100\% \longrightarrow p = f(\phi)$$

Cutting-velocity reduction ratio for various geometric progressions

ϕ	1.06	1.12	1.25	1.4	1.6	2.0
$p\,[\%]$	5.65	10.7	20	28.6	37.5	50

v_u Upper cutting velocity $\qquad v_l$ Lower cutting velocity $\qquad d$ Work diameter

Fig. 5.40 Cutting-velocity reduction ratio

various ranges of cutting diameter. By applying the speeds indicated in the diagram, the nearest possible approximation to the upper cutting velocity is obtained. When the applicable diameter range has been exceeded, the next speed must be engaged. The saw-tooth form of the v–d diagram is thus

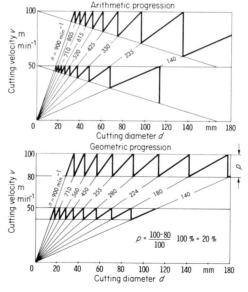

Fig. 5.41 Cutting velocities for arithmetic and geometric speed progressions

obtained for the total diameter range. (In contrast to this, the exact value of v_u may be obtained for any cutting diameter in stepless drives.)

It can be seen that the curves are fundamentally different for geometric and arithmetic progressions. In the case of arithmetic progressions, the cutting-velocity reduction ratio is dependent upon the given cutting speed as well as the speed interval c. Should the drive be designed to satisfy the upper cutting velocity at low speeds (corresponding to large cutting diameters) then the steps would be unnecessarily small for the higher speeds. On the other hand, if the design is suitable for low speeds, then the cutting-velocity reduction ratio will be too large for the lower speeds. Either way, some speed ranges will not be properly satisfied. In contrast, the cutting-velocity reduction ratio in a geometric progression is only governed by the common ratio ϕ, i.e. when changing from one speed to the next lower one the cutting-velocity reduction ratio is always constant. In Fig. 5.40 the cutting-velocity reduction ratio p is stated for various common ratios of geometric progressions. This advantage enables output speeds in a geometric progression to be obtained when such drives are arranged in series. (When such drives are used in a series connection the speeds obtained are governed by the multiplication of the common ratios ϕ or ϕ^z).

As a result of these features, which do not occur in any other stepping system, main and feed drives in machine tools are mainly of the geometric progression stepped type. Geometric progression drives are designed in accordance with the preferred numbers standard (DIN 323, BS 2045), which results in simplified designs (standardization of drives), calculations and production.

The fundamental law for preferred numbers is:

$$x = (\sqrt[R]{10})^K \tag{5.80}$$

The value of R determines the number of subdivisions within a power of ten and K determines the position of the preferred number within a series. When $R = 5$, 10, 20 and 40, we obtain the basic series R5, R10, R20 and R40 respectively.

The use of preferred numbers provides a number of special features and advantages; products, quotients, whole-number powers, decimal multiples and divisions of preferred numbers are also preferred numbers.

In machine-tool construction, the basic series R20 is mostly used in conjunction with a common step ratio $\phi = 1.12$, as shown in Fig. 5.42. By selecting every second, third, fourth, etc., preferred number the sub-series R20/2, R20/3, R20/4, etc., are obtained respectively. The common ratios of the sub-series then increase with respect to the basic series to ϕ^2, ϕ^3, ϕ^4, etc. In view of the on-load speeds of typical three-phase motors (with 6% slip, approximately 2800 or 1400 rev min^{-1}) the R20 series is most common as it contains both of these values.

To facilitate the design of stepped and stepless drives with respect to their output speed, drafting conventions have been developed which considerably

Basic series R 20	Nominal rev min⁻¹						Limiting rev min⁻¹ values of the R20 series			
	R 20/2	R 20/3 (..2800..)	R 20/4		R 20/6 (..2800..)		Due to mechanical variables		Due to mechanical and electrical variables	
			(..1400..)	(..2800..)						
φ = 1.12	φ = 1.25	φ = 1.4	φ = 1.6	φ = 1.6	φ = 2		−2%	+3%	−2%	+6%
1	2	3	4	5	6		7	8	9	10
100 112 125 140 160	112 140	11.2 16	125	1400 140	112	11.2	1400	98 110 123 138 155	103 116 130 145 163	98 110 123 138 155	106 119 133 150 168
180 200 224 250 280	180 224 280	22.4	180 250	2000 2800	180 224 280	22.4	2800	174 196 219 246 276	183 206 231 259 290	174 196 219 246 276	188 212 237 266 299
315 355 400 450 500	355 450	31.5 45	355 500	4000	355 450	45	355	310 348 390 438 491	326 365 410 460 516	310 348 390 438 491	335 376 422 473 531
560 630 710 800 900	560 710 900	63 90	710	5600 8000	560 710 900	90	5600	551 618 694 778 873	579 650 729 818 918	551 618 694 778 873	596 669 750 842 945
1000			1000				980	1030	980	1060

The series R20, R20/2 and R20/4 may be extended upwards or downwards by dividing or multiplying by 10, 100 etc.
The series R20/3 and R20/6 are shown for three powers of ten, because the numbers repeat only after every fourth power

Fig. 5.42 Rotational speed ranges for machine tools (DIN 804)

simplify diagrammatic presentation, as shown in Fig. 5.43. The 'engagement network' is the basis for drive design. From this the various engagements which are possible within the drive and on the individual gear engagements are obtained. The engagement network is usually drawn symmetrically and

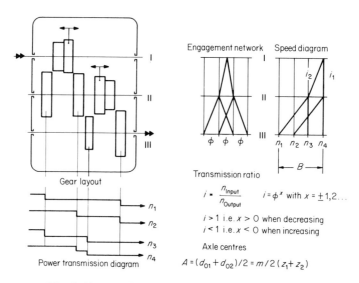

Fig. 5.43 Fundamentals for a drive design layout

each continuous line represents one transmission step. The values of the overall speed ratios cannot be read from the engagement network as they are not drawn to a speed scale.

Next to the engagement network, the speed diagram indicates the speed of each shaft and the overall speed ratio. If a geometric progression (ϕ = constant) is depicted in the speed diagram then by applying a logarithmic scale the output speed increases are represented by equal linear distances. These dimensions can also be interpreted as indicating the ratios between successive speeds as well as the exponential power of ϕ. The transmission ratios are represented by the slopes of the connecting lines for the speeds of successive shafts. For the example shown on the right of Fig. 5.43, the following relationships apply:

$$i_1 = \phi^0 = 0 \tag{5.81}$$

$$i_2 = \phi^1 = \frac{n_4}{n_3} \tag{5.82}$$

Other useful diagrammatic representations are the 'gear layout' and 'power transmission' diagrams shown in Fig. 5.43. The gear layout shows the arrangement and number of shafts, gears and clutches, where these are incorporated. The design is clarified by the use of analogous line diagrams. The power transmission diagram shows the gears which are in engagement for each individual speed selection. In addition, the power transmission diagram also shows how the individual selector forks must be set to provide a given output speed.

The theoretical mathematical analysis of the drive designs is governed by constructional limitations:

(a) Transmission ratio. The single-step ratios of main drives which have to transmit considerable mechanical power should range between 0.5 and 4:

$$0.5 \leq i \leq 4, \quad i = \frac{n_1}{n_2} \tag{5.83}$$

When the ratio $i > 4$ (step down), large total numbers of teeth will be involved as well as large centre distances between shafts with the resulting increase in overall size of the drive unit.

If the ratio $i < 0.5$ (step up), undesirable rolling conditions occur between the gear wheels and any errors in gear-tooth form will increase the auxiliary dynamic forces.

(b) Limiting numbers of teeth. To avoid the need for undercutting, the number of teeth on a gear should not be below 17 or 14 (for involute teeth having a 20° pressure angle). If a smaller number of teeth is to be used, this is only permissible if special modifications are made to the tooth form (positive pitch displacement, increase in pressure angle).

(c) Module. The strength of the gear teeth increases with the module, but as the outside diameter of the gear also increases the total space requirement of the drive becomes greater at the same time (DIN 3990).

(d) Pitch displacement. This is achieved by a radial displacement of the cutter during the gear-generation process. The modification influences the tooth form, thickness and the fillet radius. For drives with moderate speed requirements, pitch displacement will make it possible to use gears with fewer than the lower limiting number of teeth, or for a given number of teeth, make it possible to adhere to a particular centre distance of the shafts (DIN 3960).

(e) Bore diameter. The minimum distance between the root of the teeth and the bore of the gear should not be less than three times the module. In some cases the gear must be increased in size or the shaft and gear has to be manufactured in one solid piece.

(f) Centre distance of shafts. A small centre distance between the axes will reduce the size of the gears used and consequently reduce the overall size and internal inertia forces of the drive. However, for a given output torque, the loads on the teeth increase as the distance between the shafts decreases and thus the loads on the bearings also increase.

(g) Shafts. Apart from the strength requirements of the shafts, consideration must also be given to their bending and torsional distortions, which under certain circumstances could cause the effective meshing to be impaired. Apart from misalignments and reductions in the effective tooth-flank areas, even more important are the directional errors of the tooth flanks which may be introduced, leading to an uneven load distribution on the teeth with a resulting reduced load-carrying capacity.

Apart from the limitations detailed above, all components of the drive such as bearings, shafts, gears, etc., must be designed with adequate strength. To this end, a number of design techniques have been established for years, e.g. the load-carrying capacity standard DIN 3990.

Finally, a few general rules are quoted below for the design of sliding gear drives:

(a) The smallest gear of every speed step and the total number of teeth in every gear train should be as small as possible to minimize overall size and the consequential production costs.

(b) The transmission ratio should be between 0.5 and 4 and as near as possible to 1.

(c) Consideration should be given to the possibility of reducing the number of gears by compounding or looping.

(d) On multi-shaft drives the smallest transmission ratios should be as near as possible to the input of the drive, where the rotational speeds are high, the torque is low and the gear dimensions therefore are also small. The aim should be to reduce the speed at the output end of the

drive so that the resultant larger torques and gear dimensions are placed at the end of the drive unit. It is for this reason that, for example, back gears (high ratios) are situated at the end of the drive layout.
(e) Similarly, on multi-shaft drives, speed steps which are achieved in stages should have their largest factor near the input end of the drive. Thus, a three-wheel gear train will have a two-wheel train near the beginning of the drive. This will enable a large number of intermediate speeds to be obtained which will require relatively small gear sizes due to the resultant smaller torques.

5.2.2.2 Drives with stepless-speed changes
For stepless (infinitely variable)-speed changes, electrical, hydraulic and mechanical drive units are available. They are mainly used for main and feed drives. Their advantage lies in the ability to provide the exact speed demanded for any given production condition for optimum cutting or forming velocity and feed rate.

Against this, they are subject to disadvantages when compared to stepped-speed drives of lower efficiencies and higher costs.

The kinematic accuracy of stepless drives is lower than that of stepped drives. For this reason, they are currently not employed whenever the accuracy of a feed motion is of great importance, e.g. for gear generation. An improvement in the kinematic precision is possible by the introduction of a positional closed-loop control system. A tabulation of the characteristics of the various stepless drives described in the following sections (Fig. 5.44) indicates considerable differences between the differing designs which define and limit the areas of application for each of such units.

Electrical drives. The 'Ward–Leonard set' shown in Fig. 5.45 is an electric drive unit capable of producing variable-output speeds from a constant input. Its importance in recent years has somewhat diminished due to developments in power electronics (thyristor drives).

A Ward–Leonard set consists of a separately excited DC motor and an output motor (Ward–Leonard motor) in which the armature circuit is supplied with a variable voltage. To produce this variable voltage another separately excited DC machine is employed, the Ward–Leonard generator. The latter is driven by a three-phase machine at constant speed. (It is possible to use other prime movers, e.g. an internal combustion engine.)

A Ward–Leonard set may be employed in four different modes, i.e. direction of rotation and torque are reversible.[70] The speed equation of a Ward–Leonard set can be established quite simply from known relationships for DC machines (see section 5.1):

$$V_{A\,Mot} = k_{1\,Mot}\,\phi_{Mot}\,n_{Mot} + R_{A\,Mot}\,I_A \qquad (5.84)$$

$$V_{A\,Gen} = k_{1\,Gen}\,\phi_{Gen}\,n_{Gen} - R_{A\,Gen}\,I_A \qquad (5.85)$$

	Mechanical drive		Rolling drive	Switching mechanism	DC motor and controlled current rectifier
	Rubber vee belt	Chain			
Power range available (kW)	0.1...7.5 (45),	0.25...75	0.1...75	0.1...11	0.75...1000
Output speed range available (rev min^{-1})	500...4000	500...7000	0...4000	0...300	0...4000
Torque: • Increase from n_{min} to n_{max}	3.5	6	2...8	2	1
• Increase when starting	1.5	2.5	1...2.5	2.5	1
• Transmittable when stationary?	Yes	Yes	Yes	Yes	No (Yes)
Speed: • Positions available	up to 10	up to 10	∞	∞	∞
• Changable when stationary	No	No (Yes)	Yes (No)	Yes	Yes
• Accuracy of transmission (speed reduction with torque)	Medium	Low	Medium	Considerable	Considerable (low with tacho)
• No. of operating sections	4	4	4	1	1 to 4 as required
Maintenance: • Service intervals	No	2000h Oil change	2000–4000h oil change	Oil change	5000h brushes
• Dismantle for inspection?	No	No	No (Yes)	No	Yes (No)
• Typical wearing components	Belts, pulley bearings	Chain	Rolling elements discs	Relaxation oscillators	Brushes, commutators
Miscellaneous: • Efficiency pattern	Medium	Good	Medium/poor	Poor	Medium
• Noise	Medium	A little higher	A little higher	?	Medium
• Suitable for dusty, corrosive atmosphere?	Yes, with encasement	Yes	Yes	Yes	Yes, with additional protection

Fig. 5.44 Characteristics of stepless drives (single step or geared) (Ehrlenspiegel and Dittrich)

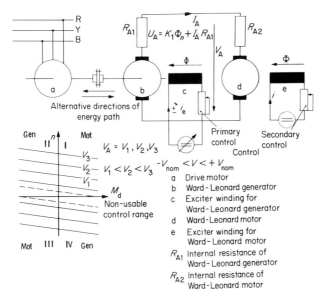

Fig. 5.45 Circuit diagram for a Ward–Leonard set

As the terminals of a Ward–Leonard generator and motor are connected to each other we have:

$$V_{A\,Mot} = V_{A\,Gen} \tag{5.86}$$

From this we obtain the following when applying equations (5.84) and (5.85):

$$n_{Mot} = \frac{1}{k_{1\,Mot}\phi_{Mot}}[k_{1\,Gen}\phi_{Gen}n_{Gen} - (R_{A\,Gen} + R_{A\,Mot})I_A] \tag{5.87}$$

The equations for the output torque T_{Mot} and power P_{Mot} are:

$$T_{Mot} = \frac{k_{1\,Mot}}{2\pi}\phi_{Mot}I_A \tag{5.88}$$

$$P_{Mot} = 2\pi n_{Mot} T_{Mot} \tag{5.89}$$

The armature current I_A is proportional to the output torque T_{Mot}. If we substitute the torque T_{Mot} from equation (5.88) for I_A in equation (5.87) we obtain the working characteristics equation:

$$n_{Mot} = \frac{1}{k_{1\,Mot}\phi_{Mot}} A \tag{5.90}$$

where

$$A = \left[k_{1\,Gen}\phi_{Gen}n_{Gen} - (R_{A\,Gen} + R_{A\,Mot})\frac{2\pi}{k_{1\,Mot}\phi_{Mot}}T_{Mot}\right]$$

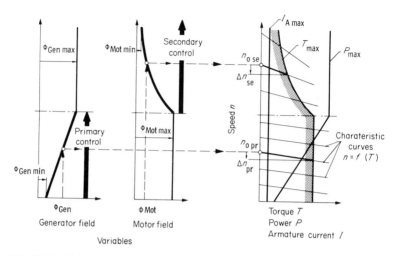

Fig. 5.46 Primary and secondary speed control of a Ward–Leonard motor

By varying either ϕ_{Gen} or ϕ_{Mot} we are able to vary the output speed n_{Mot} as shown in Fig. 5.46:

Primary control. This is motor speed control by variation of the generator field flux ϕ_{Gen}. By varying the generator field flux ϕ_{Gen} its output voltage $V_{A\,Gen}$ is varied. The motor field flux ϕ_{Mot} remains constant:

$$\phi_{Mot} = \phi_{Mot\,max} = \text{constant} \tag{5.91}$$

From equation (5.87) we note that in this case the motor speed n_{Mot} is proportional to the generator field flux ϕ_{Gen}:

$$n_{Mot} \propto \phi_{Gen} \tag{5.92}$$

When a given generator field flux ϕ_{Gen} is set we obtain initially a corresponding no-load speed $n_{0\,pr}$. As the load on the motor increases, i.e. with increasing torque T_{Mot}, the speed n_{Mot} reduces in accordance with equation (5.90) by an amount $\Delta_{n\,pr}$:

$$\Delta n_{pr} = (R_{A\,Gen} + R_{A\,Mot}) \frac{2\pi}{(k_{1\,Mot}\phi_{Mot\,max})^2} T_{Mot} \propto T_{Mot} \tag{5.93}$$

The maximum torque T_{max} is governed by the maximum permissible armature current I_{max}. This current $I_{A\,max}$ is in turn dependent upon the maximum permissible temperature rise and hence is determined by constructional and design limitations of the machine (cross-sectional area of conductors, external cooling):

$$T_{Mot\,max} = \frac{k_{1\,Mot}}{2\pi} \phi_{Mot\,max} I_{A\,max} = \text{constant} \tag{5.94}$$

$$P_{\text{Mot max}} = 2\pi n_{\text{Mot}} T_{\text{Mot max}} \propto \phi_{\text{Gen}} \tag{5.95}$$

Secondary control. This is motor speed control by variation of motor field flux ϕ_{Mot}. With secondary control the generator field flux ϕ_{Gen} remains constant:

$$\phi_{\text{Gen}} = \phi_{\text{Gen max}} = \text{constant} \tag{5.96}$$

Hence:

$$V_{\text{A Mot}} = V_{\text{A Mot max}} = \text{constant} \tag{5.97}$$

Equation (5.87) indicates that in this case the motor speed n_{Mot} is inversely proportional to the motor field flux ϕ_{Mot}:

$$n_{\text{Mot}} \propto \frac{1}{\phi_{\text{Mot}}} \tag{5.98}$$

Consequently, the speed increases when the motor field flux is reduced (weakened). Once more, when a given motor field flux ϕ_{Mot} is set a corresponding initial no-load speed $n_{0\,\text{se}}$ is obtained. In this case the speed also drops as the load increases. The amount of the decrease Δn_{se} is, however, greater when the motor field strength is reduced than that caused by primary control, because of the inverse proportionality to the motor field flux ϕ_{Mot}:

$$\Delta n_{\text{se}} = (R_{\text{A Gen}} + R_{\text{A Mot}}) \frac{2\pi}{(k_{1\,\text{Mot}} \phi_{\text{Mot}})^2} T_{\text{Mot}} \propto \frac{T_{\text{Mot}}}{\phi_{\text{Mot}}^2} \tag{5.99}$$

The maximum torque $T_{\text{Mot max}}$ is in this case proportional to the motor field flux ϕ_{Mot}:

$$T_{\text{Mot max}} = \frac{k_{1\,\text{Mot}}}{2\pi} \phi_{\text{Mot}} I_{\text{A max}} \propto \phi_{\text{Mot}} \tag{5.100}$$

$$P_{\text{Mot max}} = 2\pi n_{\text{Mot}} T_{\text{Mot max}} = \text{constant} \tag{5.101}$$

The difference between primary and secondary speed control may be seen from the different slopes of the characteristic curves and the reduction in the maximum torque T_{max}.

In the primary speed-control range the characteristic curves are steep, i.e. the speed drops only slightly as the load on the motor rises. In contrast, the curves in the secondary speed-control range are flatter, indicating that the motor is 'softer'. The speed reduces much more markedly when the load is applied. This condition is undesirable in machine tools.

The costs of a Ward–Leonard set are comparatively high, because apart from the actual drive motor two other machines are necessary. The efficiency of the complete set is approximately 65%.

Today, it is increasingly the practice to vary the speed of DC machines by direct variation of the armature voltage. This voltage variation has become possible by the availability of controlled current rectifiers (thyristors) which operate directly in the three-phase or single-phase AC circuit (see Fig. 5.7).

Hydraulic drives. Hydraulic drives transmit power by the use of an incompressible fluid, usually oil. Hydraulic drives in machine tools are almost exclusively of the hydrostatic variety.

In contrast to hydrodynamic drives, the kinetic energy of the fluid flow is of no significance in hydrostatic drives. The fluid's function is solely that of transmission of the pressure. Hydraulic drive units consist of a pump and a motor which may be placed into either an 'open' or 'closed' circuit, as shown in Fig. 5.47.

In open-circuit drives the pump obtains all the transmission fluid from the tank, whereas in a closed circuit the oil is returned to the pump from the motor, subject to any losses incurred. In a closed circuit the motor is 'hydraulically restrained', i.e. the torsional stiffness is better than that of an open circuit. Consequently, the closed circuit is suitable for braking, fast directional reversals and for feed drives when the work table tends to be 'springy', i.e. jerky feeds (e.g. due to alternating directions of cut or stick-slip effects). The advantages obtained by closed-circuit operation are paid for by considerably increased construction costs, which are not considered in Fig. 5.47. Owing to the need for heat dissipation, provision must be made for the warmed oil in

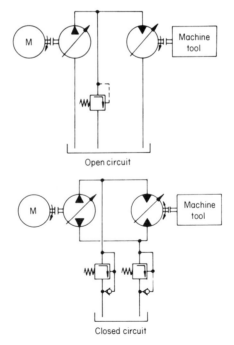

Fig. 5.47 Circuits for hydraulic drives

the circuit to be continuously exchanged with oil from the tank or be cooled by a separate cooling device. Moreover, an additional feed pump must be provided to compensate for any fluid losses incurred.

There are two methods available for variation of the motor speed. On the one hand the delivery volume of the pump may be varied (primary control) and on the other the suction volume of the motor is controlled (secondary control). The operating characteristics of hydraulic drives are similar to those of the Ward–Leonard set when these two methods of speed control are applied, which is the reason for the term 'hydraulic Ward–Leonard set' sometimes being used (compare Figs. 5.46 and 5.52).

The 'Boehringer–Sturm' drive shown in Fig. 5.48 is an internally fed vane-type pump and motor drive. The motor speed is changed by variation of the eccentricity of either the pump or motor. This drive enables the direction of the output shaft to be reversed without altering the rotation of the drive motor.

In 'radial piston drives' such as in Fig. 5.49, instead of vanes, radial pistons are employed, which are supported by the eccentrically positioned wall of the housing and are thus subjected to a linear motion. The speed variation of such drives is similar to that of vane drives. Due to better guiding and sealing of pistons and cylinders fluid losses are less than those of vane motors. The drives shown in the photographs offer the special feature of smooth-running conditions, even at low speeds.

The 'Hyvari' drive shown in Fig. 5.50 is an axial piston unit, i.e. the pistons are arranged parallel to the input and output shafts. The schematic diagram of the drive in Fig. 5.51 illustrates the method of operation. The pump (2) is connected to the input shaft (1), the motor (3) with the outer housing (5) and the output shaft (4) with the inner rotating housing (6).

The delivery volume of the pump and the suction volume of the motor is controlled by varying the inclination of both swash plates (angles α_2 and α_3). The swash plates are rigidly connected with push rods and are jointly swivelled in opposite directions. The control plate is in the centre which feeds the oil

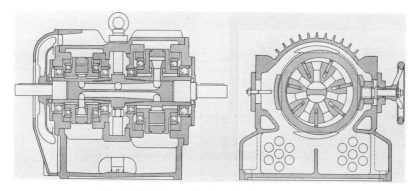

Fig. 5.48 Sturm fluid drive (Boehringer)

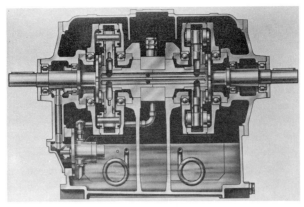

Fig. 5.49 Thomas drive (Hanauer Pumpen und Getriebebau GmbH)

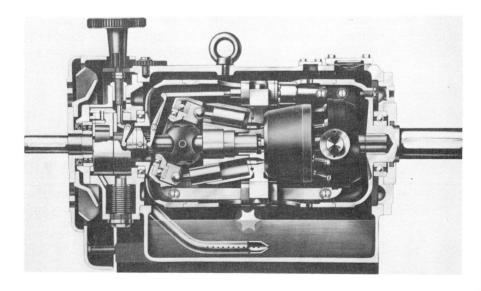

Fig. 5.50 Hyvari drive (Flender)

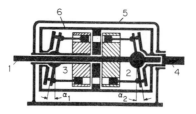

1 Input shaft
2 Pump
3 Motor
4 Output shaft
5 Outer casing
6 Inner casing (rotating)

Fig. 5.51 Schematic layout of the Hyvari drive

flow from the pump to the motor and back again. The whole of the outer casing is oil filled.

The Hyvari drive operates with an internal power fork, i.e. a part of the input energy is mechanically transmitted in the drive and the rest hydraulically. In that way, the better efficiency of the mechanical component of the transmission leads to an improved overall efficiency under certain working conditions and the bulk of the unit is reduced.

For an explanation of the way in which an internal power fork operates, let us consider the following.[71,72] If the delivery flow of a pump with a fixed housing is blocked at the output port then the pump pressure will rise until a component fractures (in practice this is not the case due to the pressure-relief valve provided). If, however, the pump casing is arranged so that it is able to rotate, then it will run at the same speed as the pump rotor. In an internal power fork a part of the fluid flow of the pump is released and absorbed by a motor with a variable suction volume. In this way, the rotational speed of the pump housing and hence the output shaft speed is infinitely variable.

Three different operating conditions may be identified:

(a) The pump swash plate is vertical ($\alpha_2 = 0$). The motor swash plate is inclined to the maximum angle. The pump does not function and idles around the inner casing. The output shaft is stationary.
(b) The motor swash plate is vertical ($\alpha_3 = 0$). The pump operates at full capacity. The torque reaction is against the stationary motor body. The output shaft is revolved by the reaction to the torque against the pump swash plate ($n_{in} = n_{out}$).
(c) The pump and motor swash plates are inclined. The power fork operates as follows:

   (1) Hydraulic energy is transmitted from the pump to the motor and a torque acts on the inner rotating casing.

(2) Mechanical energy is transmitted to the inner rotating casing as a reaction torque against the pump swash plate.

The characteristic curves of hydraulic drives (Fig. 5.52) can be presented in a similar way to those of a Ward–Leonard set. The speed control of the motor is likewise separated into primary and secondary sections. In section 5.1.2 the most important equations have already been explained:

$$\text{Pump delivery flow quantity } Q_P = n_P V_P \tag{5.102}$$

where n_P = pump speed
V_P = pump delivery volume

$$\text{Motor suction flow quantity } Q_{Mot} = n_{Mot} V_{Mot} \tag{5.103}$$

where n_{Mot} = motor speed
V_{Mot} = motor suction volume

If the fluid losses are ignored then the pump delivery flow quantity Q_P and the motor suction flow quantity Q_{Mot} in a hydraulic drive are equal. From this we obtain the motor speed n_{Mot}:

$$n_{Mot} = n_P \frac{V_P}{V_{Mot}} \tag{5.104}$$

If it is accepted that the pump oil pressure p_P is the same in the motor, then the motor torque is obtained as:

$$M_{Mot} = \frac{T_{Mot} p_P}{2\pi} \tag{5.105}$$

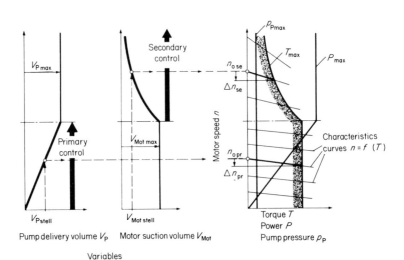

Fig. 5.52 Characteristics curves of a hydraulic drive

The motor power is proportional to the torque and speed:

$$P_{Mot} = 2\pi n_{Mot} T_{Mot} \qquad (5.106)$$

If we substitute for n_{Mot} from equation (5.104) and for T_{Mot} from equation (5.105) then the motor power P_{Mot} can be expressed as:

$$P_{Mot} = n_P V_P p_P \qquad (5.107)$$

Primary control. Motor speed control is by pump volume variation. In this case the delivery volume of the pump V_P is altered. This is achieved by adjusting the eccentricity e in the vane and radial piston pumps on the one hand and on the other by setting the angle of inclination of the swash plate α in axial piston pumps. The pump speed n_P and the motor suction volume V_{Mot} remain constant:

$$n_P = \text{constant} \qquad (5.108)$$

$$V_{Mot} = V_{Mot\,max} = \text{constant} \qquad (5.109)$$

It may be seen from equation (5.104) that in this case the motor speed n_{Mot} is proportional to the pump delivery volume V_P:

$$n_{Mot} \propto V_P \qquad (5.110)$$

When a given pump delivery volume has been selected $V_{P\,set}$ a corresponding no-load speed is obtained $n_{o\,pr}$. As the load is increased, i.e. as the torque T increases, the speed decreases owing to oil losses through leakage. The amount of the decrease Δn is in an approximately linear relationship to the torque (Fig. 5.52).

The maximum torque $T_{Mot\,max}$ is limited by the maximum permissible pump pressure $p_{P\,max}$ and the maximum suction volume of the motor $V_{Mot\,max}$, which in turn is limited by the strength of the pump and motor components and transmission lines:

$$T_{Mot\,max} = \frac{V_{Mot\,max} p_{P\,max}}{2\pi} = \text{constant} \qquad (5.111)$$

As n_P and $p_{P\,max}$ are constant, the maximum power $P_{Mot\,max}$ is proportional to the pump volume V_P:

$$P_{Mot\,max} = n_P V_P p_{P\,max} \propto V_P \qquad (5.112)$$

Secondary control. Motor speed control is by motor volume variation. By altering either the eccentricity e or the swash plate angle α the suction volume of the motor V_{Mot} is variable. In this case the pump speed n_P and pump delivery volume remain constant:

$$n_P = \text{constant} \qquad (5.113)$$

$$V_P = V_{P\,max} = \text{constant} \qquad (5.114)$$

$$Q_P = V_P n_P = Q_{P\,max} = \text{constant} \qquad (5.115)$$

From equation (5.104) it may be seen that in such a case the motor speed n_{Mot} is inversely proportional to the suction volume of the motor V_{Mot}:

$$n_{Mot} \propto \frac{1}{V_{Mot}} \tag{5.116}$$

Once again, for a given suction volume V_{Mot} a corresponding no-load speed $n_{0\,se}$ is initially obtained and the speed will again decrease as the load increases. The degree of speed reduction Δn is, however, greater than that which occurs under primary control, because the motor reacts more sensitively to leakage losses (Fig. 5.52).

The maximum torque T_{Mot} in this case is proportional to the motor suction volume V_{Mot}:

$$T_{Mot\,max} = \frac{V_{Mot} p_{P\,max}}{2\pi} \propto V_{Mot} \tag{5.117}$$

The maximum power $P_{Mot\,max}$ is constant:

$$P_{Mot\,max} = n_{P\,max} p_{P\,max} = \text{constant} \tag{5.118}$$

The costs of hydraulic drives are relatively high. The efficiency will be in the order of 80 to 90%.

Mechanical drives. Of the many mechanical stepless drives available in machine tools, belt and chain drives are mainly used. In the PIV drive (positive infinitely variable) shown in Fig. 5.53, an endless chain is placed between two pairs of coned discs. By moving the discs in the direction of their axes the

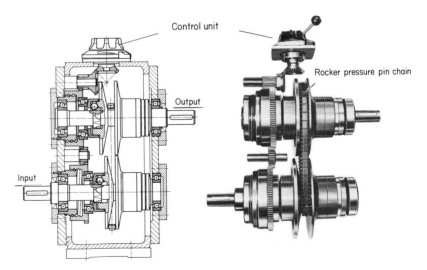

Fig. 5.53 Stepless belt drive using a power transmission chain (PIV/W. Reimers KG)

261

Fig. 5.54 Types of chain employed in PIV drives (PIV/W. Reimers KG)

distance between them is varied and a control linkage ensures that as the distance between one pair of discs is increased a corresponding reduction is obtained in the distance between the second pair of discs. This varies the contact diameters of the chain and provides the required transmission ratio.

A number of different types of chain are employed for power transmission as shown in Fig. 5.54. The 'ring roller chain' consists of steel sections which are connected to each other by rocker links. Rollers free to rotate are placed over the chain. This type of chain is suitable for velocities up to 25 m s^{-1}.

On the 'rocker pressure pin chain' the connecting pins also serve to transmit the power to the coned discs. The relatively lightweight construction of this chain permits velocities up to 30 m s^{-1}.

Each section of the 'cylinder roller chain' has two freely revolving rollers supported by each other. The manufacturers claim that a particular advantage of this form of construction is that a speed change can be effected even when the chain is stationary. Permissible chain velocities exceed 20 m s^{-1}.

Each of the above types of chain runs against plain discs which are hardened and ground. A high contact pressure is necessary to ensure a positive power transmission, as the cone angle is small.

The individual sections of the 'laminated chain' shown in Fig. 5.55 consist of sheet-metal laminations which are axially adjustable. They form teeth

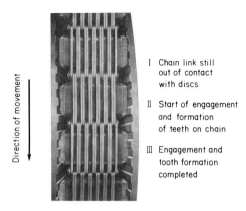

Fig. 5.55 PIV laminated chain in contact with toothed coned discs (PIV/W. Reimers KG)

which engage in slots on the coned disc providing a positive torque transmission. As the teeth re-form at each engagement, the laminations are subjected to high accelerations which limit the chain velocity to about 10 m s^{-1}. Hence slotted chains are used for powers up to about 7 kW.

The efficiency of PIV drives is in the order of 90%, which is somewhat lower than stepped gear drives.

For special applications belt drives are used in conjunction with an epicyclic gear unit, as shown in Fig. 5.56. Epicyclic gear trains[73] are a unique form of drive where planet wheels (usually three but sometimes more) mounted on a flange wheel rotate about the co-axial input and output shafts. High transmis-

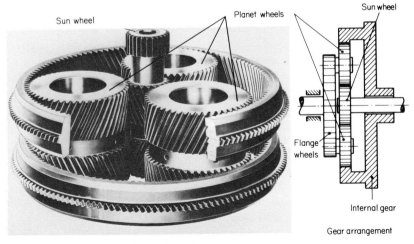

Fig. 5.56 Epicyclic gear unit

sion ratios with comparatively small space requirements can be obtained by using epicyclic gear trains.

When used in conjunction with belt drives, epicyclic gear boxes can provide drives with an external power fork as shown in Fig. 5.57. The main part of the power transmission is undertaken by the epicyclic gear unit and a smaller part by the belt drive. Two advantages are obtained from the distribution of the input torque into two or more branches within the drive and subsequent re-unification:

(a) As the power is being distributed into a number of separate paths the forces and torques acting on the various components and at the gear-engagement points are reduced. The drive unit becomes more compact and lighter when compared with a simple belt drive.
(b) It is possible to utilize within the branch which transmits the lower torque a correspondingly smaller and hence more easily manageable infinitely variable drive unit; this then provides stepless speed ratios for the complete transmission unit with a resulting increased efficiency. However, this is achieved at the cost of a reduced speed range.

For stepless drives intended for smaller power values rolling drives are more economic than steel chain units because of lower construction and manufacturing costs. The ball-disc drive shown in Fig. 5.58 is available for power transmissions of up to 3 kW. This device permits the speed to be reduced right down to zero with a constant-output torque.

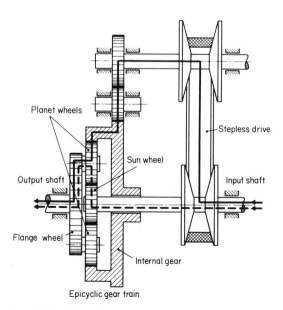

Fig. 5.57 Line of power transmission in a PIV power branching arrangement

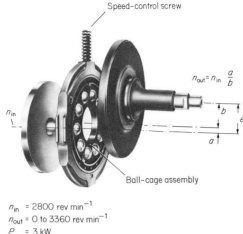

$n_{in} = 2800$ rev min^{-1}
$n_{out} = 0$ to 3360 rev min^{-1}
$P = 3$ kW

Fig. 5.58 Stepless ball/disc drive (PIV/W. Reimers KG)

Both the input and output shafts have a disc attached and their axes are displaced relative to each other by the distance e. Steel balls within a cage are placed between the plain discs which transmit the torque under pressure from one to the other. The position of the ball cage may be varied by an adjustment screw. The output speed is zero when the ball cage has its axis in line with that of the input shaft. In that position the balls idle against the output-shaft disc. If the ball cage is moved towards the centre of the output-shaft disc, then the latter will start to revolve as a result of the friction drive. As the balls take up a different path for every revolution, there is no likelihood of grooves being formed, even after prolonged running at a given setting. The pressure necessary to maintain adequate contact is produced by saucer springs.

This drive may be adjusted when stationary, the forces required for setting are low and the transmission speed is evenly maintained. When a sudden heavy load is applied, such as starting with a flywheel or sudden shock loads, the drive acts like an overload–slipping clutch as the discs will tend to slip.

Combined stepped and stepless drives. As in the case of stepped drives the overall speed ratio B, which may also be called the adjustment range in this case, is governed by the production requirements and hence by the output speeds:

$$B = \frac{n_{max}}{n_{min}} \qquad (5.119)$$

When a stepless and stepped drive are combined the overall speed ratio B is the product of the two individual speed ratios of the component drive units:

$$B = B_o B_{St} \qquad (5.120)$$

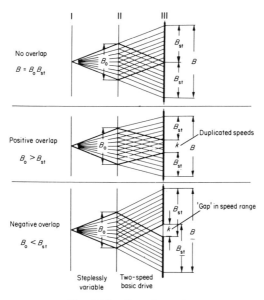

Fig. 5.59 Speed overlap

where B_o is the speed ratio of the stepless drive and B_{St} is the speed ratio of the stepped drive.

Depending upon the particular arrangement used, the terms 'positive or negative overlap' are applied. The border-line case is when the overlap is zero, as shown in the upper part of Fig. 5.59. The 'overlap ratio k' is the term applied to the ratio between the speed ratios of the stepless to the stepped drive:

$$k = \frac{B_o}{B_{St}} \tag{5.121}$$

When $k = 1$ (no overlap), B_o and B_{St} are equal and the output speeds are available without a gap in the range. When $k > 1$ (positive overlap), B_o is greater than B_{St} and a section of the speed range can be obtained through either of the steps in the stepped drive. Conversely, when $k < 1$ (negative overlap), B_o is lower than B_{St} and a speed gap ensues, representing a range of output speeds which are not available.

5.2.3 Non-uniform transmission drives

In general, when such drives are used in machine tools, a rotary motion is converted into a linear reciprocating movement for a machine component which carries either the tool or the work. With a few exceptions (e.g. the feed for external cylindrical or surface grinding) the movement is only used in one direction for the actual working process. Non-uniform drive units are also employed for auxiliary functions, such as material transportation, component handling, etc.

There is a very large number of differing transmission design principles.[74] In the following paragraphs a few of the non-uniform transmission drives are described.

5.2.3.1 Slotted-link mechanism
In shaping machines a slotted-link mechanism is employed as shown in Fig. 5.60. A sliding block is fixed to a crank which slides in the slotted link when the crank rotates, producing a linear movement for the ram. Such a drive provides a fast and a slow stroke as a result of its design. The fast stroke is used for the non-cutting return motion whereby the time required for this idle movement is reduced. The velocity diagram shown in Fig. 5.61 indicates that a constant machining velocity can only be approximated during the middle of the stroke.

5.2.3.2 Crank mechanisms
In the cutting machine-tool group the crank mechanism is used for driving slotting machines. The most common application, however, is for chipless machine tools, e.g. presses as shown in Fig. 5.62, because this type of drive offers an economic method for providing large forces. The main disadvantage is that the time required for the return stroke is equal to that of the working stroke (a half-revolution of the crank).

It can be seen from the displacement diagram in Fig. 5.63 that when the crank is at its dead-centre positions (0° and 180°) the velocity is zero and the acceleration at its maximum value. Theoretically, the ram force F_s is infinite at the 180° crank angle (knee toggle principle).

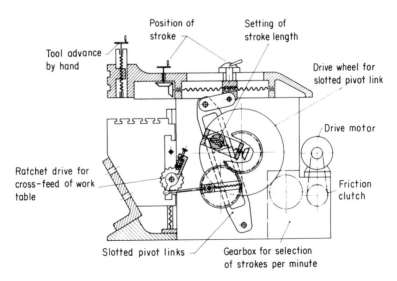

Fig. 5.60 Drive mechanism for a shaping machine (Klopp)

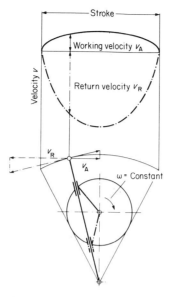

Fig. 5.61 Velocity diagrams of a mechanically driven shaping machine

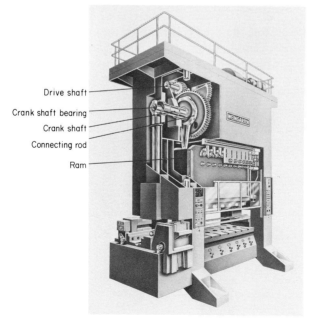

Fig. 5.62 Drive for a two-speed power press (Weingarten)

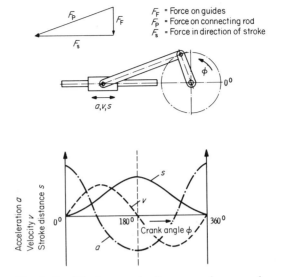

Fig. 5.63 Displacement diagram of a crank mechanism

5.2.3.3 Crank-rocker mechanism

The crank-rocker mechanism is used for the provision of the working stroke in a gear-shaping machine, as shown in Fig. 5.64. The ram in the centre of the diagram has a circular rack in the middle of its length and is moved by a gear quadrant (rocker) lever which is driven by a crank wheel. The stroke is varied by radial adjustment of the crank pin on the crank wheel. The position of the stroke is set by shortening or lengthening the connecting rod. The drive to the crank wheel is transmitted through a gear box which is fitted into the

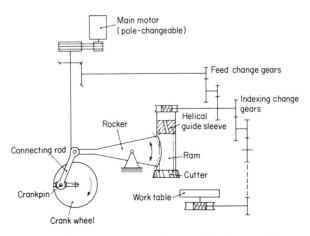

Fig. 5.64 Drive layout of a gear-shaping machine

crank housing and enables the speed to be selected from a wide range in a number of steps.

5.2.3.4 Cams

Cams enable any required motion to be provided. However, the costs of producing cams which operate without shock and backlash are high. Moreover, such drives are only suitable for transmitting small forces due to the wear on the cam surface by the follower. Hence, they are mainly used for tool and material feed motions (automatic screw machines) and for material feed devices on stamping machines and presses. (Other non-uniform drives are discussed in Volume 1 when metal-forming machines are dealt with.)

5.3 Couplings and clutches

This section is based on references 75 to 82.

5.3.1 General requirements

Shaft couplings used in main-drive power transmissions are required to undertake different duties, as may be seen from the classification based on their main characteristics shown in Fig. 5.65. In contrast to the drive units, couplings only transmit the torque but do not convert it. Two main groups are identified: permanent couplings and clutches. Permanent couplings are subdivided into rigid and flexible types, whilst clutches may be distinguished between those which are externally operated and those which are self-acting. All types are constructed to provide a positive transmission with no power losses.

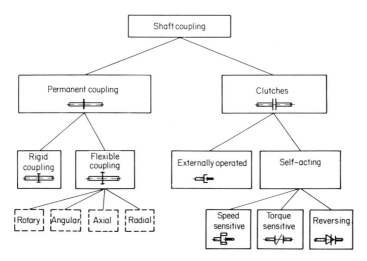

Fig. 5.65 Classification of shaft couplings (VDI 2240)[81]

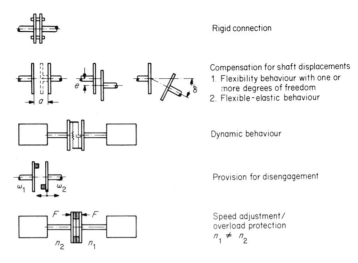

Fig. 5.66 Functions of shaft couplings

The following main features may be listed in accordance with the function to be performed by the coupling (Fig. 5.66):

(a) Ease of assembly. The connection of separate machine units by couplings into a single entity. The reason is that units are easily accessible and readily replaced.
(b) Compensation for displacement: the smoothing of alignment errors, axle displacements and closing distance gaps.
(c) Dynamic improvement. By using elastic damping elements (plastic, rubber), vibrations and impact effects are reduced.
(d) Clutch engagements/disengagements provide for the interruption of a rotating motion, i.e. separation and connection of shafts.
(e) Relative motions and engagement provide for speed matching and protect against overloading of costly drive components.

In some special cases couplings are used as heat insulators to interrupt the heat flow between two units. In the following sections a few representative designs of couplings and clutches are described.

5.3.2 Permanent couplings

Permanent couplings are used to make a connection between individual machine elements (shafts) which are intended to remain connected during service. Depending upon the individual requirement, they may be either rigid or flexible couplings. The most important features of each type of coupling are presented below.

5.3.2.1 Rigid couplings

Rigid couplings are notable for their small physical size. Whilst their cost is low, they require that the shafts are accurately aligned, which may prove expensive. Apart from the method of force transmission, differentiation may also be made between positive and friction torque transmission. The pin-and-sleeve connections shown in Fig. 5.67 are the simplest designs of positive rigid couplings. The maximum torque which may be transmitted is limited by the bearing strength of the hole and the bending and shear strengths of the pin. When higher torques have to be transmitted, flanged or clamp couplings may be used. As may be seen from the lower part of the diagram the two halves of the couplings are either connected positively with the shafts or in a combined positive and friction mode. A simple friction connection may also be made in the two couplings shown in the lower half of the diagram. These types of couplings are employed when a rigid transmission is acceptable, i.e. there is no need to take up the load with an elastic unit and it is not necessary to cater for any shaft misalignments.

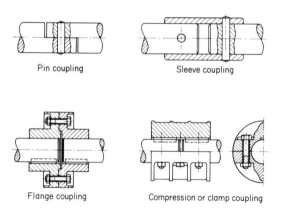

Fig. 5.67 Rigid couplings (positive)

5.3.2.2 Flexible couplings

There are two types of coupling in this category: those which permit misalignments of the shafts but have no flexibility with regard to the torque transmission and those which additionally have some elasticity in that respect. The former type of coupling facilitates the assembly and linkage of machine units and compensates for displacements of the shaft ends caused by the load application or thermal distortions.

In general, such a coupling incorporates in its construction an intermediate component which is capable of movement in a linear and angular mode but is rigid as far as torque transmission is concerned, thus absorbing any displacement between the two shafts which are to be connected. A common construction form is shown in Fig. 5.68. The curved-back-tooth gear coupling shown

Fig. 5.68 Gear-coupling with curved-back teeth
(Tacke)

has crown-formed teeth on the shaft flanges. The moving sleeve-type intermediate member has internal teeth at its ends which engage with the curved-back teeth on the flange. Such a coupling must be lubricated because there is a small relative movement of the sleeve teeth against the flange teeth with which they are in mesh during every revolution, if there is axial or angular

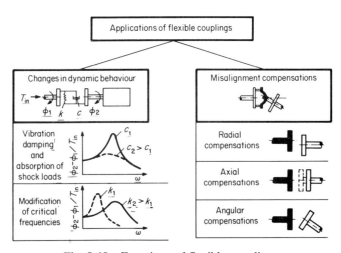

Fig. 5.69 Functions of flexible couplings

misalignment between the shafts. A curved-back-tooth gear coupling is capable of transmitting high torques.

Elastically flexible couplings are also available as standardized units. They reduce shock loads during torque transmissions and are able to absorb displacements of the shafts at the same time. A main component of such flexible couplings is an elasticated intermediate element, e.g. a rubber sleeve, rubber covered pins, etc. These reduce any torque shock which may be introduced by temporarily storing the mechanical energy (potential spring energy) whilst at the same time converting the vibration energy into heat as a result of the damping characteristics of the rubber. By varying the torsional stiffness critical natural frequencies may be altered as desired. The functions of these couplings are illustrated in Fig. 5.69.[75,76,80]

Three typical examples of elastically flexible couplings are shown in the following diagrams. Figure 5.70 illustrates a flexible-rim coupling and a

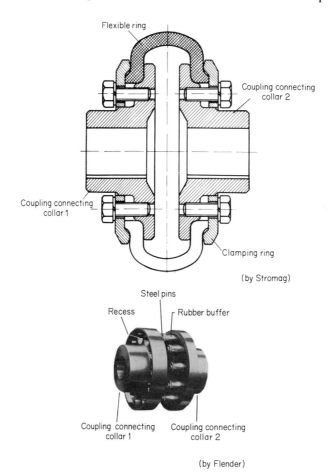

Fig. 5.70 Elastically flexible couplings

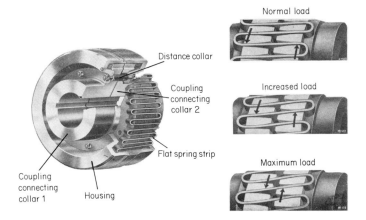

Fig. 5.71 Snake-spring couplings (Stromag)

flexible-pin coupling. The flexible element between the two halves of the coupling is made from rubber or rubber with canvas reinforcing inserts. The coupling shown in Fig. 5.71 has a snake-shaped spring-steel strip inserted between the teeth of the two halves of the coupling; the teeth are tapered and become narrower along the bending line of the spring strip towards the centre of the coupling, reducing the free length of the spring as the load increases (progressive spring characteristics). This is clarified in Fig. 5.71 for various load conditions.

5.3.3 Clutches

Clutches are classified in accordance with their method of operation. They may be engaged or disengaged 'externally' with a tripping device or alternatively the engagement or disengagement may occur due to internal centrifugal forces, torque or reversal of rotation of the clutch itself (self-acting). Below, externally operated and self-acting clutches are considered in turn in accordance with the subdivision shown in Fig. 5.65.

5.3.3.1 *Externally operated clutches*
The engagement or disengagement of the two halves of the coupling may be activated, depending upon their design, either when stationary or when running by the operation of a mechanical, hydraulic or electrical engagement device. With regard to the torque transmission, this may again be positive or by pressure being applied to friction surfaces. The externally operated couplings are of major importance in machine-tool design, and are therefore described in some detail.

Positive clutches. This type of clutch features a very simple construction whilst being capable of transmitting large torques in spite of its small overall dimen-

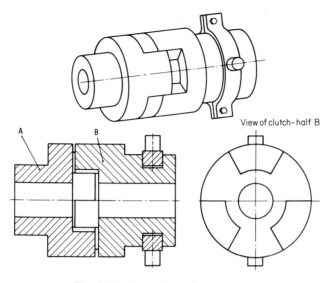

Fig. 5.72 Dog clutch (Breitbach)

sions. When disconnected there is no torque transmission at all, and engagement is only possible when stationary or when both halves are rotating at the same speed. The most common design of a positive, externally operated clutch is shown in Fig. 5.72 with the example of a dog clutch. The torque is transmitted by the positive engagement of the three jaws on each half of the clutch. The time required to engage this type of clutch is relatively long.

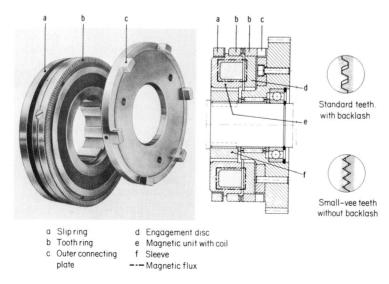

a Slip ring
b Tooth ring
c Outer connecting plate
d Engagement disc
e Magnetic unit with coil
f Sleeve
–·– Magnetic flux

Fig. 5.73 Slip-ring magnetic jaw clutch and application example (Siemens)

Another design is the slip-ring magnetic-jaw clutch shown in Fig. 5.73. This consists of a magnetic plate (e) with its coil built in, the tooth ring (b), the slip ring (a) and the engagement disc (d) with the mating teeth which transmit the torque to the other half of the clutch through the outer connecting plate (c). Both tooth rings are made from nitrided steel and are therefore wear-resistant. The moving engagement disc is pulled against the magnetic unit by the magnetic flux when the clutch is engaged. This engages the two tooth rings, enabling the torque to be transmitted. Couplings of this type may be engaged only when stationary or when the speeds of the shafts are almost synchronized. The difference in speed permissible at the point of engagement and the maximum torque which may be acting at that moment is governed by the torsional flexibility and the moment of inertia of the system. Disengagement may take place at any speed or load. The slip-ring magnetic-jaw clutch is available with two different tooth forms.

Friction clutches. A major advantage of this type of clutch is the ability to engage, usually without shock, irrespective of the difference in speeds between the shafts and whether free running or under load. However, due to the work done during engagement, heat is generated and the clutch is subject to wear. Moreover, larger overall dimensions are involved when compared with positive clutches.

Basically, the torque is transmitted when the two halves of the clutch are connected under pressure. The surface shape of the contact area may be conical, cylindrical or one or more circular rings.

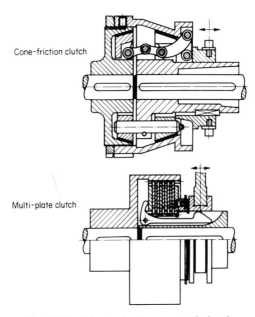

Fig. 5.74 Mechanically operated clutches

The most common designs are the cone friction clutch and the multi-plate clutch shown in Fig. 5.74. The friction force in the cone clutch is maintained either by the self-locking device of the tapers or by a continuously applied axial force. Such clutches are simple, economical, reliable and are short in overall length. Their disadvantages are that engagement is not as shock-free as in other designs and a large overall diameter is necessary.

The advantages of multi-plate clutches stem from the ability to transmit high torque values while remaining small in length and overall diameter due to the large number of friction areas. Engagement is shock-free. The disadvantages are the need for heat dissipation and the residual friction which is present in varying degrees in the disengaged state.[79]

The engagement characteristics and the load take-up conditions are now considered using a multi-plate clutch as an example.

Engagement characteristics of multi-plate clutches. The engagement characteristics and the load take-up of the clutch are presented in Fig. 5.75. Diagram (1) shows the torque characteristics of the motor T_{Mot} (three-phase squirrel-cage motor) of the multi-plate clutch and two load curves in relation to speed. In diagram (2) the clutch torque is plotted against time and in diagram (3) we can see the load take-up speed ω_{Load} and the motor speed ω_{Mot} during that time. The engagement and load take-up period can be divided into three separate phases:

Phase I: Engagement of the clutch. During this time the plates are moved towards each other and pressed together by the engagement mechanism. The clutch torque rises fairly quickly from the no-load torque T_{Free}. Depending upon the magnitude of the load

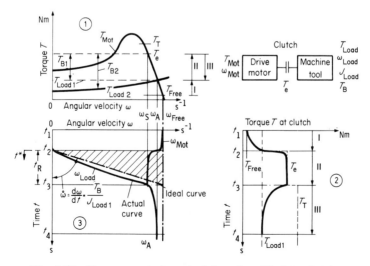

Fig. 5.75 Engagement characteristics of multi-plate clutches

being taken up, the motor speed readily drops ($\omega_{Free} \to \omega_s$). At the time t_2 the coupling torque becomes greater than the load torque T_{Load}. The output shaft begins to turn.

Phase II: Take-up of the load with slipping plates. The clutch torque reaches its maximum value, i.e. the engagement torque T_e. The magnitude of T_e, which may be considered to be almost constant, is governed by the construction of the clutch and the coefficient of friction between the plates. In this phase, the major part of the work done in acceleration is on the clutch. At the time t_3 the load speed reaches the reduced motor speed ω_s. Both halves of the clutch run at the same speed.

Phase III: Joint speed increase of motor and load. When the two halves of the coupling are running at the same speed, the acceleration of the drive is no longer governed by the clutch (T_e) but by the motor torque T_{Mot}. As a result of the rapidly dropping acceleration torque T_B, due to the steep slope of the motor curve, the joint speed rise of the load and the motor is slower. At the time t_4 the load torque is equal to the motor torque. A steady working speed ω_A is reached. The indicated torque which the clutch is transmitting during this third phase T_T is greater than the engagement torque due to the static friction between the clutch plates. At this torque load, the clutch will begin to slip.

Calculation of load take-up. The simplified calculation of the load take-up is based on conditions in Phase II. If it is assumed that the engagement torque of the coupling T_e is constant during that phase and if the speed drop of the motor is ignored, then:

$$T_B(\omega) = T_e - T_{Load}(\omega) = J_{Load} \frac{d\omega_{Load}}{dt} \tag{5.122}$$

The speed of the output shaft is given by:

$$\omega_{Load}(t) = \frac{1}{J_{Load}} \int_{t_2}^{t^*} [T_e - T_{Load}(\omega)] \, dt^* \tag{5.123}$$

For a given load condition, such as, for example, $T_{Load\,2}(\omega)$ in Fig. 5.75, the integration may be mathematically or graphically carried out (compare with section 5.1.3). In the special case where the load curve $T_{Load\,1}$ = constant the integration is simplified as:

$$\omega_{Load}(t^*) = \frac{T_e - T_{Load}}{J_{Load}} t^*$$

$$\omega_A = \frac{T_e - T_{Load}}{J_{Load}} (t_3 - t_2) \approx \omega_{Mot} \tag{5.124}$$

From this, we obtain the time during which the clutch plates slip and are rubbing against each other during the load take-up:

$$t_R = t_3 - t_2 = \frac{\omega_{Mot} J_{Load}}{T_e - T_{Load}} \tag{5.125}$$

The total work done by the clutch during this time is given by:

$$W_{tot} = T_e \int_{\phi_{Mot,t_2}}^{\phi_{Mot,t_3}} d\phi_{Mot} = T_e \phi_{Mot} \Big|_{\phi_{Mot,t_2}}^{\phi_{Mot,t_3}} \tag{5.126}$$

$$W_{tot} = T_e \omega_{Mot}(t_3 - t_2) = T_e \omega_{Mot} t_R \tag{5.127}$$

Substituting for t_F from equation (5.125) this then becomes:

$$W_{tot} = \frac{T_e}{T_e - T_{Load}} J_{Load} \omega_{Mot}^2 \tag{5.128}$$

The frictional work done during the load take-up period which is wholly converted into heat energy is:

$$W_{Frict} = T_e \int_{t_2}^{t_3} [\omega_{Mot} - \omega_{Load}(t)] \, dt \tag{5.129}$$

where $\omega_{Load}(t) = \omega_{Mot} t/(t_3 - t_2)$ (Fig. 5.75, diagram 3) and $(t_3 - t_2)$ is from equation (5.125). We have:

$$W_{Frict} = \frac{1}{2} \frac{T_e}{T_e - T_{Load}} J_{Load} \omega_{Mot}^2 \tag{5.130}$$

When comparing equations (5.128) and (5.130) it can be seen that half of the total work done by the clutch during the engagement period is converted into heat, as shown in Fig. 5.76. Consequently, if frequent engagement is expected it is necessary to provide for easy heat dissipation; otherwise the clutch plates are likely to become overheated and the oil carbonized.

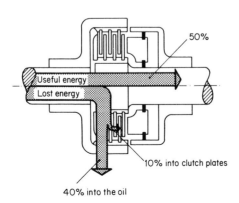

Fig. 5.76 Energy flow during engagement

Operation of multi-plate clutches. The operation of multi-plate clutches may be mechanical, hydraulic, pneumatic or electrical. For each of these methods, special design considerations apply which are considered below:

(a) Mechanically operated multi-plate clutches. In the case of these clutches, as shown in the lower half of Fig. 5.74, engagement is effected by an axial moving pressure ring. The plates are alternately fixed to the inner and outer rings of the clutch and are pressed against each other by a rocker lever, thus introducing the frictional contact.

(b) Hydraulically operated multi-plate clutches. Figure 5.77 shows such a clutch. The pressure of the piston is transmitted to the rotating plate assembly over a thrust bearing. Internal lubrication of the plates is provided in order to enable a high operating power to be engaged. Wear of the plates is compensated by the piston stroke, making clutch adjustments unnecessary.

As the working cylinder of the clutch does not rotate, the in-feed of the pressurized oil is simple and reliable. Furthermore, the pressurized oil is not subjected to any centrifugal forces and the clutch may be easily disengaged, making rapid engagements and disengagements possible.[78]

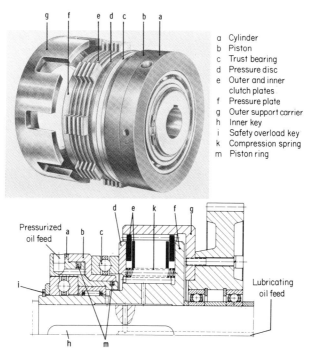

a Cylinder
b Piston
c Trust bearing
d Pressure disc
e Outer and inner clutch plates
f Pressure plate
g Outer support carrier
h Inner key
i Safety overload key
k Compression spring
m Piston ring

Fig. 5.77 Hydraulically operated multi-plate clutch and example of application (Siemens)

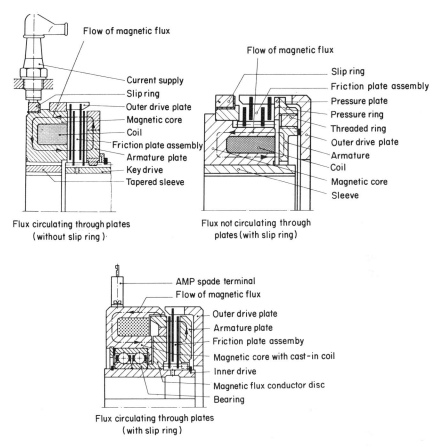

Fig. 5.78 Electrically operated multi-plate clutches (Siemens)

(c) Electrically operated multi-plate clutches. Figure 5.78 shows three designs of electromagnetically operated multi-plate clutches. On the upper left a slip ring multi-plate clutch is shown with the plates situated in the magnetic flux flow. This restricts the material choice of the clutch plates (ferromagnetic characteristics). The clutch consists of the magnetic core with a cast-in coil, the superimposed slip ring, the armature plate and the clutch-plate assembly. The outer and inner drive units for the clutch plates vary according to the particular application of the clutch. The magnetic force generated between the magnetic core and the armature plate pulls on the latter and thus presses the outer and inner clutch plates, which are situated between the core and plate, against each other. It is not necessary to provide special arrangements for adjustment to compensate for clutch-plate wear.

In the upper right of Fig. 5.78 a slip-ring multi-plate friction clutch is shown in which the magnetic flux does not pass through the clutch plates. By fitting the plates outside the magnetic-flux path a wider choice is available for the material from which they are to be made, this being a distinct advantage. The clutch-plate assembly is pressed together by a pressure plate. To ensure that this pressure is acting fully on the clutch plates, a small air gap must remain between the magnetic core and the armature plate even when the clutch is engaged. In this way any clutch wear is compensated by the air gap.

In the multi-plate clutch without slip rings, shown in the lower part of Fig. 5.78, the magnetic flux passes through the clutch plates and the current supply is not continuous. Due to the absence of slip rings, this design is also suitable for higher speeds. The maximum operating speed is governed by the highest running speed permissible on the bearing.

Examples of applications of such clutches are shown in section 5.2.2.1.

Figure 5.79 illustrates a type of clutch which operates on the induction principle. The torque transmission is effected without physical contact by the overlapping of the magnetic forces of the inner pole (electromagnet) with the magnetic field produced by the induced current in the outer pole (similar to a squirrel-cage motor). The torque characteristics may be varied through the exciter curent. There is no wear and maintenance costs are low. Such a clutch may be used as an acceleration, overload or resilient coupling. The clutch will always exhibit some slip, i.e. loss of power.

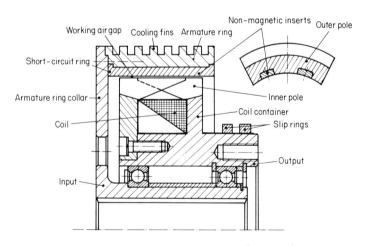

Fig. 5.79   Induction clutch (Stromag)

5.3.3.2   *Self-acting clutches*

The types of clutch which come into this category are those which engage or disengage automatically due to some limiting parameter having been reached, such as speed (centrifugal force), torque, change of direction of rotation or over-run motion, as shown in Fig. 5.65. A few typical examples in this group are described below.

Speed-sensitive clutches. A well-known speed-operated clutch is the centrifugal clutch shown in Fig. 5.80 which operates as a friction clutch. It may be operated in either direction of rotation, because the shoes are pivoted. Such a clutch is used in cases where the drive must accelerate without load and engage with the machine to be driven at higher speeds, e.g. squirrel-cage motors with a relatively low starting torque. The centrifugal force and hence the friction torque increase in proportion to n^2:

$$F = mr\omega^2 \tag{5.131}$$

where $\omega = 2\pi n$. When the strength of the return springs is increased the engagement of the friction drive takes place at a higher speed.

Torque-activated clutches. The main application of this type of clutch is as an overload protective device. When such clutches are disengaged, then according to their principle of operation they are either completely separated by a mechanism or a continuous slip occurs between their two halves. Both positive and friction drive designs are available in this type of clutch. Figure 5.81 shows two designs of positive-drive clutches in this category. The two halves of the clutch featuring a ball ratchet are pressed together by saucer springs in one case and compression springs in the other, providing a positive drive. When a predetermined limiting torque is exceeded the ball seating slips over

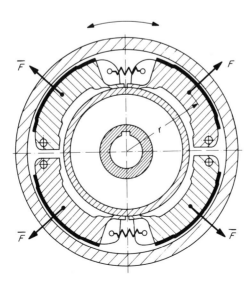

Fig. 5.80 Centrifugal clutch operating in either direction

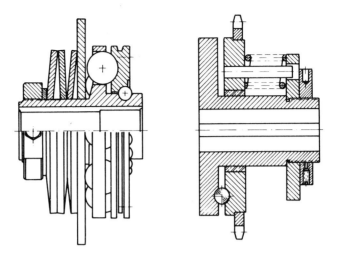

Fig. 5.81 Ball ratchet as overload clutch

the ball. This type of clutch offers the advantage of comparatively small overall dimensions.

Friction-drive torque-overload clutches are multi-plate clutches where the limiting friction torque can be adjusted through the axial pressure on the clutch plates.

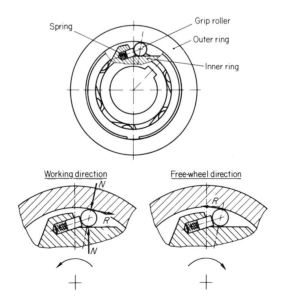

Fig. 5.82 Roller clutch grip roller free wheel (Stoeler)

Direction-activated clutches (free-wheel over-run). Such clutches have a freewheeling and working direction, i.e. in one direction there is a torque-free relative movement between the two halves of the clutch (over-run or free-wheel direction). When the direction of rotation is reversed, the clutch automatically engages and transmits the torque. This makes a special engagement device unnecessary. As an example of this type, Fig. 5.82 shows a grip roller and expanding friction clutch. This is a positive-drive direction-operated clutch. In the upper part of Fig. 5.82 the construction is shown. The rollers are pressed against the outer ring by spring-loaded pins. In the working direction shown in the lower left of the diagram the rollers are clamped between the two halves of the clutch. In the free-wheel direction shown in the lower right the spring-loaded pins give way and consequently there is no torque transmission.

6

SUMMARY

This volume, sub-titled *Construction and Mathematical Analysis*, describes the requisite design criteria and the methods available for mathematical analysis for the manufacture of machine-tool components and construction units. The highest demands are made upon these machine parts because machine tools must be able to guarantee the required working accuracies and performance under static and dynamic as well as thermal stresses.

The aim of this treatment is to provide the designer with the necessary assistance for the mathematical analysis and design of the machine-tool elements and units based upon the most up-to-date knowledge of this technology and to suggest alternative solutions.

With this in mind, the machine frames and frame components, foundations, spindle bearings, guiding systems, motors for main drives, stepped and stepless drives and clutches are dealt with. An important feature is the presentation of the mathematical techniques to determine the static, dynamic and thermal behaviour.

7

REFERENCES

1. M. Weck, W. Missen and B. Thurat: Programme für Grossrechenanlagen—Programmbeschreibung, Anwendungsbeispiele, Technische Hochschule Aachen (Dec. 1976).
2. M. Weck and N. Diekhans: Übersicht Tischrechnerprogramme—Kurzbeschreibungen, Anwendungsbeispiele, Technische Hochschule Aachen (July 1977).
3. W. F. Dreyer: Über die Steifigkeit von Werkzeugmaschinengestellen und vergleichende Untersuchung au Modellen, Dissertation, Technische Hochschule Aachen (1966).
4. A. Heimann: Anwendung der Methode finiter Elemente bei der Berechnung und Auslegung von Gestellbauteilen, Dissertation, Technische Hochschule Aachen (1977).
5. M. Weck and A. Heimann: Rechnerische Analyse des statischen Verhaltens von Gestellbauteilen zur Bestimmung geeigneter Rechenmodelle und zur Erstellung eines Katalogs berechneter und bewerteter Lösungsbeispiele für Gestelle, Bericht des Landesamts für Forschung NRW, Düsseldorf (1977).
6. F. Koenigsberger: *Berechnungen, Konstruktionsgrundlagen und Bauelemente spanender Werkzeugmaschinen*, Springer–Verlag, Berlin, Göttingen, Heidelberg (1961).
7. R. Plock: Untersuchung und Berechnung des elastostatischen Verhaltens von ebenen Mehrschraubenverbindungen, Dissertation, Technische Hochschule Aachen (1972).
8. K. Beckenbauer: Möglichkeiten zur Verringerung der Ratterneigung an spanenden Werkzeugmaschinen, VDW-Konstrukteur-Arbeitstagung, Aachen (1971).
9. W. H. Groth: Die Dämpfung in verspannten Fugen und Arbeitsführungen von Werkzeugmaschinen, Dissertation, Technische Hochschule Aachen (1972).
10. P. de Haas: Thermisches Verhalten von Werkzeugmaschinen unter besonderer Berücksichtigung von Kompensationsmöglichkeiten, Dissertation, TU Berlin (1975).
11. M. Weck *et al.*: Anwendung von Digitalrechnerprogrammen zur Berechnung von Maschinenteilen nach der Methode finiter Elemente, VDW-Konstrukteur-Arbeitstagung, Aachen (1975).
12. R. Noppen: Berechnung der Elastizitätseigenschaften von Maschinenbauteilen nach der Methode finiter Elemente, Dissertation, Technische Hochschule Aachen (1973).
13. K. E. Buck, D. W. Scharpf, E. Stein and W. Wunderlich: *Finite Elemente in der Statik*, Verlag von Wilhelm Ernst & Sohn, Berlin, Munich, Düsseldorf (1973).
14. R. Finke: Berechnung des dynamischen Verhaltens von Werkzeugmaschinen, Dissertation, Technische Hochschule Aachen (1977).

15. L. Zangs: Berechnung des thermischen Verhaltens von Werkzeugmaschinen, Dissertation, Technische Hochschule Aachen (1975).
16. M. Weck, W. Melder, W. Brey, M. Klöcker and N. Roschin: Geräuschemission von Drehmaschinen. Forschungsbericht Nr. 181 im Auftrag des Bundesministers für Arbeit und Sozialordnung und des Vereins Deutscher Werkzeugmaschinenfabriken e.V., Bundesanstalt für Arbeitsschutz und Unfallorschung Wirtschaftsverlag NW, Bremerhaven.
17. M. Weck, W. Melder, W. Brey, M. Köcker and N. Roschin: Geräuschemission von Fräsmaschinen. Forschungsbericht Nr. 214 im Auftrag des Bundesministers für Arbeit und Sozialordnung und des Vereins Deutscher Werkzeugmaschinenfabriken e.v., Bundesanstalt für Arbeitsshutz und Unfallforschung, Wirtschaftsverlag NW, Bremerhaven.
18. M. Weck, W. Melder, W. Brey, M. Klöcker and W. Wiedeking: Geräuschemission spanender Werkzeugmaschinen—Wälzfräsmaschinen, Kreissägemaschinen, Schleifmaschinen und Bohrmaschinen. Forschungsbericht Nr. 264 im Auftrag des Bundesministers für Arbeit und Sozialordnung und des Vereins Werkzeugmaschinenfabriken e.v., Bundesanstalt für Arbeitsschutz und Unfallforschung, Wirtschaftsverlag NW, Bremerhaven.
19. W. König and M. Weck: Zahnrad- und Getriebeuntersuchungen. Bericht über die 21. Arbeitstagung am 21. und 22. Mai 1980, Laboratorium für Werkzeugmaschinen und Betriebslehre der RWTH, Aachen.
20. VDI 2567: Schallschutz durch Schalldämpfer, Verein Deutscher Ingenieure (Sept. 1971).
21. M. Weck and W. Melder: *Maschinengeräusche—Messen, Beurteilen,* Mindern, VDI-Verlag, Düsseldorf (1980).
22. VDI 2057 (Draft Parts 1 to 3): Assessment of the effects of mechanical vibrations on human beings, Deutscher Normenausschuss (Jan. 1979).
23. M. Heckl and H. A. Müller: *Taschenbuch der Technischen Akustrik,* Springer-Verlag, Berlin, Heidelberg, New York (1975).
24. VDI 2711: Noise reduction by enclosures, Verein Deutscher Ingenieure (June 1978).
25. VDI 3720, Parts 1 to 3: Noise abatement by design, Verein Deutscher Ingenieure (Oct. 1975, July 1977 and April 1978).
26. M. Weck, W. Melder, W. Brey and M. Klöcker: Larmminderung an spanenden Werkzeugmaschinen. Katalog geräuschmindernder Massnahmen. Forschungsbericht im Auftrag des Bundesministers für Arbeit und Sozialordnung und des Vereins Deutscher Werkzeugmaschinenfabriken e.V, Sonderschrift Nr. 8, Bundesanstalt für Arbeitsschutz und Unfallforschung, Wirtschaftsverlag NW, Bremerhaven.
27. R. Westphal: Untersuchung zur Geräuschentstehung und Geräusch–minderung an Kreissägeblättern für die Leichtmetallbearbeitung, Dissertation, TU Hannover (1979).
28. M. Weck and K. Teipel: *Dynamisches Verhalten spanender Werkzeugmaschinen,* Springer-Verlag, Berlin, Heidelberg (1977).
29. VDI 2062, Parts 1 and 2: Shock and vibration isolation, Verein Deutscher Ingenieure (Jan. 1976).
30. Bundes-Immissionsschutzgesetz (Federal Imission Protection Law): Bundesgesetzblatt I, pp. 721–743, *Bundesanzeiger* (March 1974).
31. DIN 4150 (Tentative Standard) Parts 1 to 3: Vibrations in buildings, Deutscher Normenausschuss (Sept. 1975).
32. B. Thurat: Maschine-Fundament-Baugrund, Bestimmung des Gesamtverhaltens bei statischer und dynamischer Belastung, gezeigt am Beispiel von Werkzeugmaschinen, Dissertation, Technische Hochschule Aachen (1978).
33. M. Weck and B. Thurat: Projektstudie 'Werkzeugmaschinenfundamente', Technische Hochschule Aachen (1977).

34. D. Fuller: *Theorie und Praxis der Schmierung*, Berliner Union, Stuttgart (1960).
35. A. Kingsbury: On problems in the theory of fluid-film lubrication with an experimental method of solutions, *Trans. Am. Soc. Mech. Eng.*, **53**, 59–75 (1931).
36. S. J. Needs: Effects of side leakage in 120 degree centrally supported journal bearings, *Trans. Am. Soc. Mech. Eng.*, **56**, 721–732 (1934); **57**, 135–138 (1935).
37. DIN 50281: Friction in bearings: Beuth–Verlag, Berlin and Cologne (1960).
38. H. B. Bongartz: Die Tragkraftkomponenten der Gleitführung und ihr Einfluss auf das Reibungsverhalten, Dissertation, Technische Hochschule Aachen (1970).
39. E. Saljé: *Elemente der spanenden Werkzeugmaschinen*, Carl Hanser–Verlag, Munich (1968).
40. M. B. Dolbey: The normal dynamic characteristics of machine tool plain slideways, Dissertation, University of Manchester (1969).
41. G. Kretschmer: Die Führungsgenauigkeit von Mehrgleitflächenlagern für Werkzeugmaschinenspindeln der Feinbearbeitung, *Maschinenbautechnik*, **17**, No. 8, 415–418 (1968).
42. H. Peeken: Der Einsatz von mehrflächigen Gleitlagern in Werkzeugmaschinen höchster Genauigkeitsansprüche, *Werkstatt und Betrieb*, **92**, No. 10, 729–738 (1959).
43. VDI 2201: Design of bearings: introduction to the performance of slide bearings, Beuth–Verlag, Berlin and Cologne (1968).
44. VDI 2204: Gleitlagerberechnung; Hydrodynamische Gleitlager für stationäre Balastung, Beuth–Verlag, Berlin, and Cologne (1968).
45. G. Vogelpohl: *Betreibssichere Gleitlager—Berechnungsverfahren für Konstruktion und Betrieb*, Springer–Verlag, Berlin, Göttingen, Heidelberg (1958).
46. E. Schmid and R. Weber: *Gleitlager*, Springer–Verlag, Berlin, Göttingen, Heidelberg (1953).
47. H. Opitz: Aufbau und Auslegung hydrostatischer Lager und Führungen und konstruktive Gesichtspunkte bei der Gestaltung von Spindellagerungen mit Wälzlagern, VDW-Konstrukteur-Arbeitstagung, Technische Hochschule Aachen (Febr. 1969).
48. H. Opitz: Ausführung und Anwendung hydrostatischer Lager und Führungen im Werkzeugmaschinenbau—Bericht über das 13. Aachener Werkzeugmaschinen-Kolloquium. *Industrie-Anzeiger*, **90**, No. 67, 1527–1536 (1968).
49. G. Porsch: Über die Steifigkeit hydrostatischer Führungen unter besonderer Berücksichtigung eines Umgriffs, Dissertation, Technische Hochschule Aachen (1969).
50. R. Böttcher: Untersuchungen über das dynamische Verhalten hydrostatischer Spindellagerungen, Dissertation, Technische Hochschule Aachen (1968).
51. W. Effenberger: Hydrostatische Lageregelung zur genauen Führung von Werkzeugmaschinenschlitten, Dissertation, Technische Hochschule Aachen (1970).
52. W. Miessen: Berechnung des statischen und dynamischen Verhaltens hydrostatischer Spindel-Lagersysteme, Dissertation, Technische Hochschule Aachen (1973).
53. G. Porsch: Ausführung und Untersuchung einer hydrostatischen Spindelmutter, *Industrie-Anzeiger*, **91**, No. 5, 97–100 (1969).
54. H. Opitz: *Moderne Produktionstechnik, Stand und Tendenzen*, 242–264, Verlag W. Girardet, Essen (1970).
55. A. Wiemer: *Luftlagerung*, VEB Verlag Technik, Berlin (1969).
56. E. Blondeel, R. Snoeys and L. Devrieze: Aerostatic bearings with infinite stiffness, *Annals of the CIRP*, **25**, No. 1 (1976).
57. J. Schmidt: Berechnung und Untersuchung aerostatischer Lager aus porösem Werkstoff, Dissertation, Universität Karlsruhe (1972).
58. Catalogues of bearing manufacturers.

59. E. Wiche: *Die radiale Federung von Wälzlagern bei beliebiger Lagerluft*, SKF Kugellagerfabriken, Schweinfurt, WTS 93, 1.
60. H. Pittroff and E. Wiche: *Laufgüte von Werkzeugmaschinenspindeln*, SKF Kugellagerfabriken Schweinfurt, WTS 690820 (1969).
61. M. Weck: Rechnerunterstützte Auslegung und Auswahl von Lagern, Unpublished report, Laboratorium für Werkzeugmaschinen der RWTH, Aachen (1978).
62. L. Lachonius: Kreiselmomente an Wälzlagern, SKF Kugellagerfabriken, Schweinfurt, *SKF-Zeitschrift*, No. 136 (1963).
63. M. Weyand: Hauptspindellagerungen von Werkzeugmaschinen: Reibungs- und Temperaturverhalten der Wälzlager, Dissertation, Technische Hochschule Aachen (1969).
64. K. Honrath: Über die Starrheit von Werkzeugmaschinenspindeln und deren Lagerungen, Dissertation, Technische Hochschule Aachen (1960).
65. P. Vanherck: Dimensioning of liquid film dampers, CRIF—Report MC 26, Leuven, Belgium (Nov. 1968).
66. T. Bödefeldt and H. Sequenz: *Elektrische Maschinen*, Springer–Verlag, Vienna (1971).
67. *Dubbels Taschenbuch für den Maschinenbau*, Springer–Verlag, Berlin, Göttingen, Heidelberg (1974).
68. Hütte, Vol. II A, B. Verlag Wilhelm Ernst & Sohn, Berlin (1960–1963).
69. *Betriebshütte, Vol. Fertigungsmaschinen*, Verlag Wilhelm Ernst & Sohn, Berlin (1964).
70. J. W. Schroeder: Grundzüge der elektrischen Maschinen und Antriebe für Maschineningenieure, Reprint of lecture, Technische Hochschule Aachen.
71. W. Backé: Grundlagen der Ölhydraulik, Reprint of lecture, Technische Hochschule Aachen.
72. I. N. Bronstein and K. A. Semendjajew: *Taschenbuch der Mathematik*, Verlag Harri Deutsch, Zürich and Frankfurt/Main (1968).
73. J. Looman: *Zahnradgetriebe*, Konstruktionsbücher Band 26, Springer–Verlag, Berlin, Heidelberg, New York (1970).
74. P. Grodzinski and G. Lechner: *Getriebelehre I, Geometrische Grundlagen*, Sammlung Göschen, B. 1061, Verlag Walter de Gruyter & Co, Berlin (1960).
75. G. Niemann: *Maschinenelemente*, Vol. I, Springer–Verlag, Berlin, Göttingen, Heidelberg (1960).
76. D. H. Bruins: *Werkzeuge und Werkzeugmaschinen*, Part 1, Carl Hanser–Verlag, Munich (1968).
77. G. Köhler and H. Rögnitz: *Maschinenteile*, Part 2, B. G. Teubner Verlagsgesellschaft, Stuttgart (1965).
78. L. Ernst and W. Rüggen: Eine neue, hydraulisch geschaltete Lamellenkupplung, *Industrie-Anzeiger*, 93, No. 99, 2495–2498 (1971).
79. H. Hasselgruber: Der Einrückvorgang von Reibungskupplungen, *Forsch. Ing.-Wes.*, 20, No. 4, 120–125 (1954).
80. W. Beitz: Untersuchung der elastischen und dämpfenden Eigenschaften drehelastischer Kupplungen und ihrer Dauerfestigkeit, Dissertation, TU Berlin (1961).
81. VDI 2240: Systematische Einteilung der Wellenkupplungen nach ihren Eigenschaften, Beuth–Verlag, Berlin and Cologne (1971).
82. G. Spur: *Spanende Werkzeugmaschinen*, Sonderdruck 1–12 aus ZWF, Carl Hanser–Verlag, Munich (1975–1976).

INDEX

Absorption ratio, 67
Acoustic power readings, 62–63
Air consumption, 169
Air hammering, 171
Amplitude–frequency characteristic, 76
Apertures, 21–22
Asynchronous machines, 216–220
  construction and working principles, 216–217
  types of rotors, 218–220
Axial piston motors, 225–226

Back gearing arrangement, 237
Bearing play, 178
Bearing resilience, 179
Bearing slackness, 178
Bearings, 11, 88–205
  aerodynamic and aerostatic, 167–173
  characteristics of, 91
  circular radial, 159
  comparison between radial and axial resilience curves, 184–185
  design variations, 125–127
  double-row rolling, 186
  dynamic behaviour, 198–201
  filmatic, 126
  function of, 88
  grinding-machine spindle, 127
  grinding-wheel spindle, 128
  hydrodynamic multi-face, 124
  hydrodynamic plain, 92–129
  hydrodynamic plain circular, 121–129
  hydrodynamic spindle units, 127
  hydrodynamic thrust, 127
  hydrostatic axial, 160
  hydrostatic linear, 150–157
  hydrostatic plain, 129–167
  hydrostatic plain circular, 157–163
  hydrostatic spindle, 165
  main spindle, 90
  multi-face, 128
  precision boring-machine spindles, 127
  pre-loaded, 179–185, 197, 200
  pre-loaded tapered roller, 197
  pressure build-up and acceleration characteristics, 122–125
  radial, 158, 179–183
  rolling, 185–186, 189–203
  rolling cylindrical, 176–190
  self-adjusting for temperature variations, 125
  spindle units, 190–203
  static behaviour, 193–198
  thermal behaviour, 201–203
  three-face, 125
  thrust, 127, 183–184
  types of, 91
Belt drives, 240, 262–263
Bending resistance, 14, 17, 29
Boehringer–Sturm drive, 255
Bolted joints, 25–28
Boring machines
  horizontal, 11, 12
  vertical, 56, 57, 196
Boundary, conditions, 53
Bredt formula, 15

Cage, slip, 185
Cams, 269
Capillary resistors, 134, 140, 142, 145, 162
Caro-expansions bearing, 126
Chain drives, 260–261
Circular-saw cutters, 72
Clamping devices, 117–119
Closed-loop drive, 238–239, 241
Clutches, 274–285
  applications, 282
  centrifugal, 283
  cone friction, 277
  direction-activated, 285
  dog, 275
  electrically operated multi-plate, 281
  engagement characteristics, 277
  externally operated, 274
  friction, 276

291

Clutches (*continued*)
  friction-drive torque-overload, 284
  hydraulically operated multi-plate, 280
  induction, 282
  mechanically operated, 276
  mechanically operated multi-plate, 280
  multi-plate, 277
  overload protective, 283–284
  positive, 274
  self-acting, 282
  slip-ring magnetic jaw, 275, 276
  speed-sensitive, 283
  torque-activated, 283
Columns, 17, 20, 22, 28, 29, 43, 61
Compatibility conditions, 50
Computer aided design, 3–5
Computer calculations, 1, 29, 43
Computer programs, 3–5, 54–55
Couplings, 269–274
  classification of shaft, 269
  flexible, 271–274
  functions of, 270
  gear, 272
  general requirements, 269–270
  permanent, 270
  rigid, 271
  snake-spring, 274
Crank mechanisms, 266
Crank–rocker mechanism, 268
Cutting-velocity reduction ratio, 243–244

Damping, 9, 32–37, 39, 62, 66–68, 73, 120, 121, 146, 201
  fluid, 120
  friction, 120
Decision making, 1
Deflection, 18
Deformation, 9, 11–12, 22, 27
Degrees of freedom, 88, 89
Design aids, 1, 3, 4
Design considerations, 13, 33–35, 42–45
Design cycle, 1, 2
Design evaluation, 2
Design examples, 28–29
Design index, 2, 4–5, 21
Design problems, 4
Design requirements, 1
Diaphragm absorption units, 147
Difference equations, 45
Differential coefficients, 45
DIN 4150, 81
Direct current (DC) machines, 207–214
  armature control, 211

basic and operational equations, 208
construction and working principles, 207
current rectification, 213–214
field control, 211–213
general design and circuit diagram, 208
load limitations, 209
self-cooling, 212–213
speed control, 209
Direct current motors, start-up conditions, 233–234
Distortion, 15, 22, 24, 86
Distortion analysis, 59
Drive components design strength, 248
Drive-design analysis, 247
Drive-design layout, 246
Dynamic behaviour, 35, 56, 59, 76
Dynamic loading, 30–32
Dynamic quantifying factors, 32–33
Dynamic viscosity, 92, 93, 94

Economic factors, 9
Elastic deformations theory, 45
Elasticated support, 86
Electric circuit analogy, 160
Electrical drives, variable-output speeds, 249
Electrical machines, 207–220
Energy consumption, 148–149
Engagement network, 246
Epicyclic gear trains, 262–263
Equilibrium condition, 52
Ergonomics, 3

Finite element method, 18, 29, 45, 87
  application of, 46
  calculation examples, 55–62
  calculation systems, 54
  computer programs, 55
  element types, 46–47
  fundamentals of, 46–53
  survey of calculations, 53–55
Flange designs, 24
Flange thickness, 25
Flexibility, 9, 11, 19, 20, 24
Flexibility amplitude, 32
Flexibility–frequency characteristic, 32
Flexibility–frequency response curve, 58
Floor requirements, 76
Floor vibrations, 77
Flow equations, 15
Flow velocity, 96
Fluid wedge, 94–102
Force deformation relationship, 28, 33

Force diagram, 125
Force flux, 11–12, 24
Force transmission, 22
Forging hammer, 79, 82, 84
Foundation block, 75, 84–86
Foundations, 74–87
  design, 74
  machines with adequate inherent rigidity, 76–78
  medium and heavy machine tools without adequate inherent rigidity, 85–87
  metal-forming machines, 78–84
  precision machines, 75
  without adequate inherent rigidity, 84–85
Fourier analysis, 81
Frame component evaluation, 21
Frame component interfaces, 24
Frame components, 6–9
Frame construction, 7
Frame materials, 6–9
Frame requirements, 6
Friction
  boundary, 103
  fundamentals of, 92–111
  rolling, 103
  rolling/sliding, 103
  sliding, 102
  solid, 103
  types of, 102–103
Friction coefficient, 39
Friction damper, 37–39
Frictional energy, 39

Gear box, 240
Gear drive units, 237
Gear layout, 247
Gear motors, 223
Gear shaping machines, 154, 155, 268
Gear teeth
  limiting numbers of, 247
  strength of, 248
Geared headstock, 29
Grinding machines, roller, 84
Guide inaccuracies, 156
Guides
  hydrostatic, 154
  hydrostatic ram, 155
  rolling linear, 173–176
Guideway design features, 90, 111–117
Guideway elements, 111–117
Guideway surface pressure and positioning, 110–111

Guideways, 22, 88–205
  dove-tail, 113
  flat, 112
  function, 88, 90
  hydrodynamic plain, 92–129
  hydrostatic, 104, 129–167
  hydrostatic machine table, 151
  narrow, 111–112
  plastic coatings, 116
  plastic faced, 117
  pre-loading, 174–176
  roller, 104, 173–176
  setting, 88
  steel inserted, 116
  vee/flat and double-vee, 112–113
  working, 88, 89
  worn or damaged, 115
Guiding errors, 119–120
Gyroscopic couple, 184

Hagen Poiseuille law, 132, 136
Half-speed whirl, 170
Heat sinks, 60
Heat sources, 40–42, 60
  external, 39
  internal, 39–40
Hole effects, 22
Hydraulic circuits, 149
Hydraulic displacement machine, 221–222
Hydraulic drives
  characteristic curves, 258
  closed-circuit, 254
  open-circuit, 254
  speed control, 259
  speed variation, 255
Hydraulic motors, 220–233
  components of, 220–221
  linear drives, 221, 228–229
  principle of, 221
  rotary drives, 221, 223–227
  speed control, 229–233
  speed-control characteristic relationships, 229–233
  speed-control methods, 221
  swept volume per revolution, 221
Hydraulic table drive, 228–229
Hydraulic Ward–Leonard set, 255, 258
Hydrodynamic pressure build-up, 94–102
Hydrostatic drives, 221, 254
Hysteresis losses, 120
Hyvari drive, 255–257

Installation, 74–87
  criterion for, 75
  machines with adequate inherent rigidity, 76–78
  metal-forming machines, 78–84
  methods, 74
  procedures, 74

Johnson drive, 166–167
Joining methods, 24–28
Joint design, 24–25
Joint length, 20

Kinematic viscosity, 94
Kirchhoff network analogy, 130

Lathes
  inclined bed, 40, 43, 44, 114
  roll-turning, 31
Lead-screws and nuts, 165–167
Load-carrying capacity, 169
Load-dependent resistors, 135
Load-displacement conditions, 141
Load-displacement diagram, 140, 143, 145, 162, 163
Lubrication, 129, 187
  boundary, 120
  fundamentals of, 92–111
Lubrication medium, 90

Machine-bed designs, 20
Machine-bed loading, 19
Machine-bed rib designs, 18
Machine-tool construction, 3, 8
Mackensen bearing, 126
Main drive units, 206–290
Mass distribution, 32
Mass effect, 32
Mass ratio, 37
Material pairing for sliding components, 107
Material properties, 8–9
Mathematical analysis, 45–62
Mathematical model, 55, 57, 60, 61, 85, 86
Mechanical drives, 260–264
Milling cutter, 38
Milling machines
  gantry, 33, 85
  universal, 114, 240
  vertical, 241
Motor speed control, 252–253
Motors, 206–234

Mounting elements, 77–79
Mountings, 74
  natural frequency, 78

Newtonian fluids, 93
Noise, 3
Noise emission, 62–63, 70
Noise generation, 63, 64
Noise reduction, 62–73
  active, 69
    primary measures, 70
    secondary measures, 72
  examples of, 68–73
  passive, 69
    primary measures, 72–73
Noise transmission, 69
Non-linear stiffness relationship, 138
Non-Newtonian fluids, 93
Numerical control, 4

Ohm's law analogy, 130
Oil supply system, 133–136, 139, 141, 142, 166, 188

Perception strength curves, 81
Phase relationship, 33, 34
Pilot valves, 228
Piston-displacement units, 225
Pitch displacement, 248
PIV drives, 260, 262
Polar moment of area, 13
Power press, 267
Power transmission chain, 260
Power transmission diagram, 247
Preferred numbers, 245
Pressure distribution, 131, 168
Pressure gradients, 97, 122, 159
Principle of virtual work, 49

Radial piston drives, 255
Radial piston motors, 227
Radiation factor, 66
Radiation rate, 66
Recirculating ball spindles and nuts, 203–205
Rectifiers, 213–214
Resilience amplitude, 34
Resilience ratio, 181
Resonance, 147, 199
Resonance frequency, 83
Ribbing, 15–21, 29
Ribbing designs, 17, 18

Rigidity, 9, 76
Rigidity matrix for push-pull truss, 47–50

Safety requirements, 3
Seals and sealing, 163
Second moment of area, 13, 14
Shaft centre distance, 248
Shaft strength requirements, 248
Shaping machines
  drive mechanism, 266
  mechanically driven, 267
Slides 11, 22
  hydrostatic machine-tool, 153
Slideways, 29
  aerodynamic and aerostatic, 167–173
  boundary lubrication, 106
  friction characteristics, 104
  hydrodynamic, 111–121
Sliding gear drives, 248
Sliding surface preparation, 107
Slip-ring rotor motor, 218
Slotted link mechanism, 266
Soap bubble analogy, 15
Soil vibrations, 79
Solid sound barriers, 65
Solid sound damping, 65
Sound absorption, 67–68
Sound barrier ratio, 68
Sound energy, 65, 66
Sound generation, 63
Sound pressure, 65
Sound transmission, 65, 66
Speed change range, 242, 243
Speed diagram, 247
Speed ranges, 246
Speed ratios, 241–243
Speed selection, 241, 247
Spindles, 11
Spring rigidity, 37
Spring stiffness, 78
Squirrel-cage motor, 219
Star-delta switching system, 217
Start-up conditions, 233–234
Static behaviour, 55
Static loading criteria, 9
Static quantifying factors, 9–11
Stepless ball/disc drive, 264
Stepless drives, 245, 249–265
  combined stepped and, 264–265
  electrical, 249–254
  hydraulic, 254–250
  mechanical, 260–264

Stepped drive calculations, 241–249
Stepped drives, 236, 245
Stick-slip effect, 104–106, 116
Stiffness, 9–11, 18, 20–22, 25–27, 32, 33, 76, 85, 100, 101, 138, 140, 142, 144, 145, 153, 161, 199
Stiffness matrix, 49, 50, 51, 53
Stribeck curve, 103–106
Structural design, *see* Design
Swash plate, 226–227, 230
Synchronous machines, 214–216
  application of, 216
  construction and working principles, 214
  variation of rotational speed, 215–216

Temperature distribution, 42, 59
Temperature effects, 61, 187
Temperature patterns, 60
Tension–compression truss, 47–50
Thermal behaviour, 42, 59–62
Thermal deformation, 40–41
Thermal efficiency chain, 40, 42
Thermal equivalent force, 60
Thermal loading criteria, 39–45
Thermal quantifying factor, 41–42
Thermoelastic behaviour, 43
Thermoelastic deformation, 42–44
Thermoelastic distortion, 60
Thomas drive, 256
Time-dependent temperature pattern, 60
Torque loading, 15, 18, 21, 24
Torque resistance, 14–15, 29
Torsional deformation, 15, 16
Torsional resistance, 16, 17
Torsional shear stress gradient, 14
Torsional stiffness, 13, 15, 22
Transmission drives, 235–269
  general requirements, 235–236
  non-uniform, 235, 236, 265–269
  uniform, 235–265
Transmission factor, 65
Transmission impedance, 65
Transmission ratio, 247
Tremor evaluation, 82
Tremor factor, 79, 81
Truss elements, 46

Value analysis, 18
Vane-type motors, 224
Velocity distribution, 97
Vibration, 3, 9, 30–39, 76, 77, 79, 81, 146, 148, 171, 185–186

Vibration absorbers, 37, 38
Vibration damper, 37
Vibration patterns, 58
Vibrators, 83
Viscosity, 92

Ward–Leonard set, 249, 251, 253
Wear factors, 106–111
Wedge space, 94–102, 122